O9-AHT-633

ALSO BY ROBERT PETER GALE

Final Warning: The Legacy of Chernobyl
(with Thomas Hauser)

ALSO BY ERIC LAX

Faith, Interrupted

Conversations with Woody Allen

The Mold in Dr. Florey's Coat

Bogart (with A. M. Sperber)

Woody Allen, A Biography

Life and Death on 10 West

On Being Funny

RADIATION

RADIATION

WHAT IT IS, WHAT YOU NEED TO KNOW

o

ROBERT PETER GALE, M.D., PH.D.,

AND **ERIC LAX**

Alfred A. Knopf New York

THIS IS A BORZOI BOOK
PUBLISHED BY ALFRED A. KNOPF

Comments on benefit-to-risk ratios of voluntary radiation exposures are general in nature and reflect the best judgment of the authors and the many medical and scientific experts we consulted. They are meant as a guide and should not be used to make individual decisions. There are important and unresolved (perhaps unresolvable) differences of opinion among experts on several issues fundamental to assessing benefit and risk. Scientists' knowledge of the effects of radiation is evolving. Persons concerned with a radiation-related health issue should consult their physician and/or other health professionals. The U.S. and European governments and the Health Physics Society have important, unbiased useful information on this issue. Links can be found on our website, www.radiationbook.com.

ISBN 978-1-62490-716-6

Jacket design by Archie Ferguson

Manufactured in the United States of America

To my esteemed colleagues in the United States, Russia, Ukraine, Belarus, Brazil, Japan, China, and elsewhere who collaborated in response to these global nuclear and radiation accidents over three decades. They taught me much. Also, my deep respect and admiration to the many heroes, some of whom we were fortunate to save, others not, who battled and/or were victims of these tragic events.

RPG

For Jonathan Segal
Who improves writing and friendship

EL

CONTENTS

A NOTE TO THE READER

Virtually everything is radioactive, including us; some things are just more radioactive than others. Radiation comes in several forms, but it can be conveniently divided into two types: *ionizing,* which can cause cancer and other harmful effects as well as beneficial effects, and *nonionizing,* which, except for some ultraviolet radiations, generally causes little harm but gives substantial benefits. Radiation also can be naturally occurring or man made. The radiation that Americans receive is almost equally divided between both sources. Naturally occurring radiation levels are similar for Americans and Europeans, but Europeans receive considerably smaller man-made doses because they make less extensive use of medical procedures using ionizing radiations. People living in other parts of the world, such as South America, Africa, and Asia (with the exception of Japan, which has a high use of medical radiation), receive even less man-made radiation, for the same reasons. Radiation can kill you, or it can cure you.

When an accident involving radiation occurs, like the catastrophic one in 1986 at the Chernobyl nuclear power facility in the former Soviet Union and the incident in 2011 at the Fukushima Daiichi power plant in Japan, people everywhere ask a deceptively simple question: What is my risk, and my family's, from the released radiation? It is normal to worry about what harm might be carried through the air and what effects it will have on the food we give our children, on the water we drink, on sea life, and on the environment. These worries are compounded by news accounts that are sometimes contradictory and that often quote "experts" whose opinions vary wildly.

People have concerns not only about accidents at nuclear power facilities but also about many other sources of radiation: X-rays of our teeth, our chest, an injured hand or leg; CT (computed tomography) scans of our chest or abdomen; sunlight and tanning booths; and radio waves from cell phones. We want to know at what point radioactivity released from a nuclear accident raises our risk of cancer, what damage radiation can cause a pregnant woman and her fetus, and the effects of radioactivity in the atmosphere from nuclear weapons tests.

The often conflicting information about the possible health consequences of radiation from all sources is likely to leave anyone confused and wondering what they can do to decrease their risk of harm from radiation. How does your risk of cancer from radiation compare to other cancer risks in your life? These legitimate concerns deserve direct, intelligent, credible answers, which we intend to provide.

Because radiation is ubiquitous—it continually touches our lives in many ways—you will not finish reading these pages with a definitive answer to everything about radiation. That is simply impossible. But you will end up with sufficient knowledge and understanding of the subject to help you make sensible, informed decisions about radiation's health effects and risks. Although radiation is associated with many dangers, things may not be nearly as scary as you imagine. In fact, radiation saves lives every day. Americium-241 makes many smoke detectors work; tritium stimulates the phosphors that illuminate some exit signs; gamma rays test the structural integrity of airplanes, bridges, and skyscrapers; and cobalt-60 and other sources of radiation are used to diagnose and treat cancers. Frightening as the topic of this book might be in some ways, information and education can relieve some, if not most, anxiety about radiation.

RADIATION

INTRODUCTION

"THE CESIUM BOMB"

In 1985 two radiologists in Goiânia, Brazil, who treated people with cancer with radiation therapy machines containing cobalt-60 and cesium-137, moved to a new office. The physicians planned to take the devices with them, but the owner of their old clinic building claimed that the cesium-137 machine was his and held on to it. The dispute moved to the courts. Over the next year, the owner could not attract other physicians as tenants. The building remained empty, eventually fell into disrepair, and then was partially demolished. Two years later one of the radiologists returned to remove the cesium-137 unit but was stopped by police who had been called into the disagreement. The radiologist warned the owner that someone needed to take responsibility for what could happen with what he called "the cesium bomb," and he wrote to the director of the Institute for Civil Servants, who had called the police to stop him from removing the machine, alerting him to the potential radiation hazard. The court's response was to post a security guard at the facility twenty-four hours a day to deter people, especially would-be scavengers, from entering the building. Their presence worked—for four months, anyway.

On September 13, 1987, the daytime guard called in sick so he could go to a movie theater with his family and watch *Herbie Goes Bananas.* No replacement guard was sent. Two scavengers who had long heard rumors that there was valuable equipment in the building seized the opportunity to enter and saw the huge metal-encased cesium-137 device, which they assumed had value as scrap. When such a machine is intact, the cesium-137 is encased in a tungsten-and-steel capsule surrounded by lead, thus shielding anyone who comes close to it from exposure to the gamma rays that cesium-137 emits. (Someone standing three to six feet in front of such an unshielded source for one to two hours or less could receive a lethal dose of radiation.) The scavengers, unaware of the potential danger of their trophy, spent hours removing the shiny stainless-steel-cased rotating assembly containing the cesium-137, which looked like the most valuable part. It was certainly the most dangerous. The instant they removed the assembly, they were potentially exposed to the cesium-137 beam, as they would have been had the machine been turned on.

They put the assembly into a wheelbarrow, took it about a mile to one of their homes, and placed it under a mango tree in the garden. Because the assembly was no longer shielding the gamma rays emitted by the cesium-137, within forty-eight hours both scavengers suffered from dizziness, vomiting, and diarrhea. They went to a clinic and were told they had either a food allergy or food poisoning. Despite feeling ill, one of the scavengers continued trying to break open the capsule holding the cesium-137, convinced there was something even more valuable inside. Finally he punctured the thick glass window of the orange-sized capsule that held the cesium and scooped some out. He first assumed that the cesium was gunpowder, but he and a couple of coworkers could not ignite it. In touching it, however, they became contaminated; the effects would soon show up as radiation burns on their bodies.

Several days after stealing the device, one of the scavengers sold the dismantled parts to the owner of a scrap yard, who placed them in his home garage. This was when the danger escalated. That night the scrap dealer noticed a blue light emanating from his new acquisition. (The word "cesium" is derived from the Latin *caesius,* meaning "heavenly blue." The blue in this instance was fluorescence of cesium rather than radiation itself.) He was immediately enamored of the glowing powder, which he thought was valuable or perhaps even supernatural, and took it into his house to show his family. Over the next three days, he invited friends and other family members to see this wondrous substance and gave them some of it. Not only were they contaminated, but they spread cesium-137 wherever they went afterward. One man took enough of the cesium to paint a cross on his abdomen and carried the rest home to show his family. His six-year-old daughter smeared cesium on her body, proudly showed her mother how she glowed, and then swallowed some that was on her hands when she ate. Contamination was spread further when the scrap dealer sold the remnants of the machine to a second dealer.

Fifteen days after the scavengers stole the machine, the wife of the scrap dealer who had bought the device from them realized that many people in her family and among her friends were getting sick. She went to the scrap yard where the remaining pieces of the radiation device were stored, put them in a plastic bag, and took a bus across town to a medical clinic, leaving a trail of radioactive cesium-137 and potentially exposing thousands of people along the way. At the clinic she placed the bag on the desk of the doctor who saw her and told him that the contents were making her family ill. The doctor, thinking she had a tropical disease, sent her to a hospital, where some of the others who had been in contact with the cesium-137 had already been admitted, with the same diagnosis. The doctor was at first content to leave the bag on his desk, but he then grew apprehensive that it might be dangerous and moved it to a courtyard, where it remained for a day.

One of the physicians at the hospital suspected that so many people with similar skin lesions might have been harmed by radiation. He contacted the Goiânia State Environmental Protection Agency and proposed that a medical physicist look at the items in the bag. Fortunately, such a physicist was visiting Goiânia, and the next day he borrowed a Geiger counter, used for geological measurements, from a government agency and set out for the hospital. He found the readings en route so high, he assumed the device was broken. He returned for a replacement and kept it turned off as he traveled to thc hospital.

During the physicist's absence to obtain the Geiger counter, the hospital physician grew more concerned about the bag's contents and called the fire department. The physicist arrived just in time to stop the firefighters from throwing the bag into the river. After he turned on the replacement detector, he was stunned to find that whatever was in the bag was emitting radiation millions of times above normal.

News of an accident involving radioactivity spread mass confusion and worry. Authorities alerted the hospitals where the radiation victims had been and tried to find everyone who might have been exposed so that the spread of radiation could be stopped; they confiscated clothes from those known to have touched the cesium-137. The round-up eventually led to the monitoring of more than 110,000 people, many of whom were brought to the town's soccer stadium for evaluation and triage. There they were given showers to decontaminate themselves and housed in tents. The easiest way to determine who in a large group is in most need of help after a radiation accident is to ask that everyone who has become nauseated to take one step forward. Nausea implies a dose of at least 1,000 millisieverts (more on these soon); this is the level at which, after two days, blood cells affected by radiation die, and complications like anemia, bleeding, and infection arise.

Here the story takes another turn. The Brazilian navy had a secret nuclear program designed to counter a perceived Argentinean effort to build a nuclear weapon, and officials worried that news of a radiation accident might scuttle their program. They gathered employees of a commercial nuclear power facility—the only people in the country with extensive radiation experience—and flew them to Goiânia. There they used Geiger counters to check the thousands of people in the football stadium for radioactive contamination. In all, 249 people were determined to have been in contact with cesium-137. One hundred twenty had slight radioactive residue on their skin or clothing and were quickly decontaminated by thorough washing. The remaining 129 required greater attention: 79 had skin or external exposure that required treatment but not hospitalization, and 50 showed a higher level of exposure; 20 of them were admitted to a hospital. Bone marrow failure (which halts the production of blood cells and is fatal if not reversed) developed in 14, and 10 of them and 4 other patients were secretly transported by plane to Marcílio Dias Naval Hospital (Hospital Naval Marcílio Dias) in Rio de Janeiro.

The doctors in Rio had little experience with a radiation emergency of this magnitude. However, Dr. Daniel Tabak, a hematologist, had worked with one of us. Bob Gale a year earlier had helped treat firefighters and others who received very high doses of radiation when they responded to the explosion and fires at the Chernobyl nuclear power facility. (He has participated extensively in treating victims of almost every major nuclear accident in the past twenty-five years and in assessing the long-term health implications of those accidents, including the one at Fukushima.) Tabak tracked Bob down in Bonn, Germany, where he had just spoken to a parliamentary committee about nuclear issues, and asked if he would come immediately to Rio.

As soon as Bob heard what had happened, he knew—from his experience with the Chernobyl victims and from his work with his colleague David Golde at UCLA—that recently developed hormones called recombinant human granulocyte-macrophage colony-stimulating factor (rHuGM-CSF), which were then in clinical trials on people receiving anticancer chemotherapy, would be useful. Bob and his Soviet colleagues had used this drug at the Chernobyl accident (more on this later), and he knew it was not available in Brazil. It stimulates the bone marrow cells to produce granulocytes, the white blood cells that fight infections. The bone marrow of people with severe radiation sickness cannot produce sufficient blood cells to keep them alive, so doctors transfuse them with red blood cells (which carry oxygen) and with platelets (which clot blood), along with antibiotics and antivirus drugs. Granulocytes, however, cannot be effectively given by transfusion—they must be generated within the body.

Bob called Roland Mertelsmann, a colleague in Frankfurt am Main, a hundred miles away, who was testing the hormones. Mertelsmann and Sandoz, the Swiss pharmaceutical company with which he was working, agreed to give Bob some for use in Brazil. He sped to Frankfurt, collected the drug, packed in dry ice in a Styrofoam box, and barely made the last flight of the day.

Bob arrived in Rio with no visa and a box steaming carbon dioxide from the evaporating dry ice. The lack of a visa was not a problem, nor was the vaporous package. But his visibility was. A year after the Chernobyl disaster, Bob was still easily recognizable. The Brazilian navy didn't want a doctor who was famous for treating radiation victims to be seen in Rio. Tabak met him, whisked him through immigration and customs, then had him lie down in the backseat of a car so that he would not be seen as they drove to a hotel and then the navy hospital.

Because the victims had absorbed cesium-137 by handling it and even by eating and drinking it, their bodies were radioactive (the cesium-137 in the person was radioactive, not the person) and thus were a risk to those who cared for them. No pregnant nurses or nurses of childbearing age were allowed onto the medical team because of the potential radiation damage to their unborn children. To avoid any unnecessary exposure to the radiation emitted from the victims, the doctors and nurses operated behind lead shields. That, however, proved impractical. These were acutely ill people, and it was impossible to care for them with such encumbrance. Bob and the others accepted that they would be exposed to radiation that had a low risk of causing them harm later in life. Fortunately, none have shown ill effects in the ensuing twenty-five years.

Four of the eight people who received the bone marrow hormone survived. The four who died included the wife of the original scrap dealer who had taken the bag of parts to the health center and the young girl who ate and smeared herself with the cesium-137. Radiation killed their white blood cells, which allowed bacteria to take over, and they died of their infections. One victim needed to have a forearm amputated because of severe radiation burns. But ten of the fourteen victims taken to Rio survived their ordeal, as did all those treated in Goiânia. (Bob had a much smaller drama. The admiral of the Brazilian navy in charge of the secret program liked him very much but worried that Bob might expose the program after he left the country, half-jokingly confiscated his passport just before his departure, then returned it.)

A key lesson from this story is that not being aware of the inherent dangers of radiation and radioactivity can prove harmful or even fatal. But another lesson is that the dangers of radiation are not necessarily what you suppose them to be. Sometimes an event that seems to have all the makings of a catastrophe capable of harming hundreds or thousands of people actually harms relatively few. There often is a great difference between what we fear and what is real, and that is a gap we hope to close.

RADIATION AND YOU

Earth, born more than 4.5 billion years ago, is a radioactive planet in the radioactive solar system in the radioactive universe made by the Big Bang, which happened 9 billion years earlier. Radiation is older than the universe—thorium-232 has a half-life of about 14 billion years, almost three times longer than the age of Earth—yet we have known about it only since 1895, when the German physicist Wilhelm Conrad Röntgen (1845–1923) discovered X-rays. Without radiation, there would be no life on Earth.

All of us are radioactive. We, and our environment, exist by virtue of green plants that capture photons, the basic units of light energy—they are produced by thermonuclear fusion within the Sun. Plants use these photons via photosynthesis to separate water into hydrogen and oxygen. The hydrogen is then combined with carbon dioxide from the atmosphere to produce glucose, which the plant burns to produce energy. The oxygen is released into the atmosphere, and we and virtually all other living creatures breathe it. (There are some organisms that exist without oxygen.) Energy produced by burning glucose is transferred to us when we eat plants, or animals that feed on plants or plant products.

Subatomic particles and electromagnetic waves of radiation flow from the Sun through the universe and collide with Earth (and with us). Photons have the properties of both a particle and a wave but they have no mass, and they and the subatomic particles are much smaller than we can see or even imagine. Electromagnetic waves are categorized not by size but by their frequency (which is the opposite of their wavelength—the distance between two waves) and by their energy, which is proportional to their frequency: if you look at a line showing repeated waves, one wavelength is the distance from the peak of one to the peak of the next. Some waves are so close together that their wavelength is a thousand times smaller than the head of a pin. Other waves are so far apart that their wavelength is longer than half a mile. The shorter the distance between waves, the greater their energy and the greater their potential to harm. When waves or particles from the Sun or other sources strike matter (like us), their energy, or at least some of it, is transferred to that matter. The greater the energy transferred and the greater the radiation dose, the greater the danger to humans. The scale from the least dangerous waves (those with the longest wavelengths) to the most dangerous (those with the shortest) has radio waves at one end and then microwaves, infrared radiation, visible light, ultraviolet (UV) radiation, X-rays, and gamma rays at the other end. The difference between them is vast. An X-ray delivers about 10 million times more energy than a radio wave, which gives a sense of why radio waves don't hurt us but X-rays can.

Atoms and subatomic particles are at the core of our existence, yet they are astonishingly small. To have a sense of just how small, let's say a grapefruit is full of nitrogen atoms. If you made each of those atoms the size of a blueberry, the grapefruit would then have to be the size of Earth to hold them. For you to be able to see the nucleus of one atom, a blueberry would have to be the size of a football stadium; the nucleus would be the size of a small marble. It stretches the imagination to grasp just how dense a nucleus is, but try this: mentally, take 6 billion or so cars and squeeze them into a box 1 foot by 1 foot by 1 foot. And that's just one atom. Even smaller than the nucleus are subatomic particles like protons and neutrons (which give the nucleus most of its mass) and electrons.

Electrons and high-speed electrons (also called beta particles) are atomic particles that are *fundamental particles*, meaning they cannot be broken into smaller parts (although some recent experiments suggest they can). Neutrons are also particles (composed of quarks, which are also fundamental particles). In contrast to protons (also composed of quarks), which have a positive electrical charge, neutrons have no charge. Two protons and two neutrons that stick together and have mass make up an alpha particle.

All three types of particles—electrons, neutrons, and alpha particles—can harm humans. Neutrons are the most dangerous because they are very energetic, can penetrate deeply, and deposit large amounts of energy into tissues.

Alpha particles are intermediate in their danger to humans because they are comparatively large and they deposit all their energy locally, they do not penetrate as deeply as neutrons. Electrons are the least dangerous because they cannot penetrate very deeply and because they have relatively less energy to deposit into matter. The illustration on page 14 shows that electrons penetrate deeper than alpha particles, but as a particle moves through matter, an alpha particle deposits much more energy than does an electron. When an atom releases alpha particles, the process can also result in the release of gamma rays (high-energy electromagnetic radiation), which can also harm humans. Sometimes these particles, especially alpha particles, can enter the body in a place where they can cause considerable local damage. For example, people living in areas where there are high concentrations of radon gas inhale large numbers of radon atoms into their lungs, which then release alpha particles into the surrounding tissue. This is an important cause of lung cancer, especially in nonsmokers.

Electromagnetic waves cannot be seen, with one very important exception: visible light. (See the first illustration in the insert.) Visible light, though, is a *very* small part—less than 1 billion-billionth—of the electromagnetic spectrum, all parts of which are forms of radiation (energy in motion).

We live on a planet that is 93 million miles from the Sun—not too near, not too far, just right for our oxygen-and-carbon-based life. Life on Earth has evolved to exist in the conditions the planet offers. If we were a paltry 50,000 or 100,000 miles closer to or farther away from the Sun, the photons we rely on to sustain life would be either too strong or too weak for life as we know it.

We may think of ourselves as beings of thought and limb, of flesh and blood, but actually we are all atoms and mole-

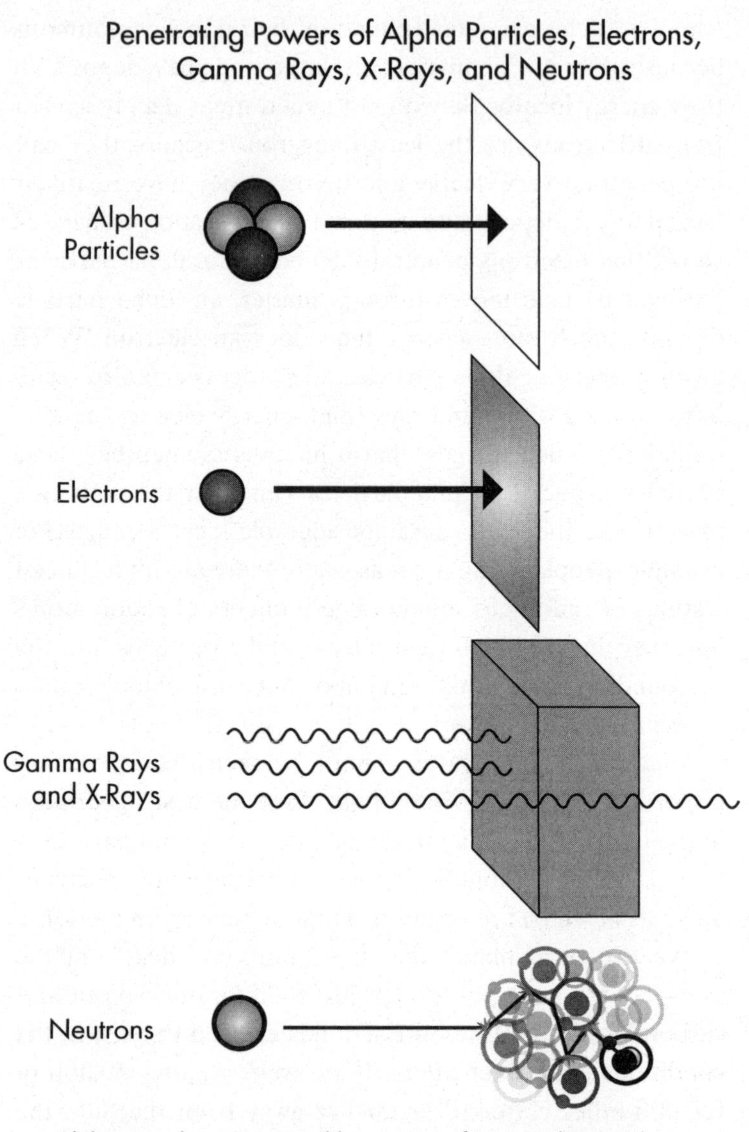

Alpha particles are stopped by a piece of paper, electrons by a layer of clothing or a thin piece of aluminum, gamma and X-rays by several feet of concrete or a few inches of lead, and neutrons by concrete or several feet of water.

cules undergoing constant chemical reactions, trillions upon trillions of them every second, encased in the semipermeable membrane that is our skin. These reactions power our brain, our heart, our muscles, and our vision. They lead to growth of old cells or, when we exercise or need to repair an injury, to the birth of new cells; they make us age, and eventually, when they run out of steam, we die. When our chemistry changes, so do we. Cells that perform a function start doing it differently or not at all. Chemical changes in cells can be advantageous; for instance, radiation from the Sun, in the form of ultraviolet rays (UVB), triggers cells in the skin to produce vitamin D from cholesterol. Vitamin D is essential to help us absorb calcium from our diet and maintain our bony skeleton. Something like plants, we capture photons from the sun. But UVB rays can also be dangerous—they can cause mutations in the DNA of skin cells that can end in cancer. Our chemistry is our destiny.

Many forms of radiation, like microwaves and radio waves, have insufficient energy to cause important changes in the cells they strike. But other, more energetic forms can alter the structure of atoms they hit, forcing electrons out and producing a charged particle—an ion. These forms of radiation are termed *ionizing radiations* and they are everywhere. Some are produced by the natural decay of radionuclides (atoms that are radioactive) remaining from the creation of the universe. Others are man made or man caused, coming from exploding nuclear weapons, burning coal (which releases naturally occurring radionuclides locked inside raw coal), fissioning uranium in nuclear power facilities, and many other sources. Ionizing radiation in sufficient quantities can be a life changer.

Heat is a form of energy. The amount of heat it takes to turn a normal cell into a cancerous one is roughly that contained in a cup of hot chocolate. The difference is that the energy in the hot chocolate is broadly diffused and so warms the whole cup, whereas the energy in ionizing radiation is as focused as a firmly struck cue ball is to a rack of billiard balls. Just as the energy from the cue ball knocks apart the rack, ionizing radiation literally knocks an electron out of an atom and produces ions that can magnify radiation damage.

Atoms contain a nucleus surrounded by a cloud of electrons. Not all nuclei are stable. Unstable nuclei undergo radioactive decay, releasing the ionizing radiations we have been speaking about. Sometimes this nuclear instability is the result of a nucleus that absorbed a subatomic particle. Unstable nuclei can emit combinations of gamma rays, electrons, and subatomic particles. Rather as a wet dog shakes off water, the nucleus wants to be rid of what is not normal for it. In thermodynamic terms, it seeks the most stable configuration. The period of instability varies from substance to substance at a known rate called a *half-life*—the time it takes for one-half the radioactivity to be emitted. It can last from nanoseconds (copernicium-285) to billions of years (thorium-232, uranium-238). (We will go into the details in chapter 3.) Ionizing radiation can somctimes prevent cells from doing their work as designed. For example, when radiation damages DNA, it is usually repaired quickly. But if the repair is improper, the chemical changes can lead to a cancer, another illness, or death.

Radioactivity is in our food and our water, even in our bodies. It is part of our makeup and causes no known harm. There are several radioactive elements in each of us, among them naturally occurring radioactive forms of potassium (potassium-40) and carbon (carbon-14), as well as man-made isotopes like cesium-137 that result from nuclear fission. Each second, thousands of radioactive atoms in our bodies decay; sleep next to someone, and your bedmate will get a dose of radiation from you. Potassium-40, which to the body looks like normal nonradioactive potassium, is taken up by all cells but especially muscle cells. Men, who usually have more muscle mass than women, are on the whole more radioactive than women because they have more potassium-40.

About half of the radiation we normally receive comes from natural sources called *background radiation.* There are two major sources of background radiation: *cosmic radiation,* which comes from the universe, including our Sun (cosmic radiation increases when there are solar flares) and supernovas (that fling out particles when they explode); and *terrestrial radiation,* which comes from radionuclides in the Earth's crust. An additional component comes from radiation in our body. We live in a sea of radiation. So when scientists want to analyze radioactivity in a sample of something (or us), they have to shield their radiation detectors with lead or other dense material to block out background radiation. Because the explosion of atomic bombs in the atmosphere released radionuclides that never before existed on Earth, objects made after 1945 contain man-made radioisotopes. In contrast, steel made before 1945 is less radioactive than that manufactured afterward and is prized for making radiation detector shields.

Radon-222, an odorless, colorless gas, is a decay product of radium-226 and a link in the uranium-238 and thorium-232 decay chain that ends in lead-206, which is not radioactive. It is everywhere on Earth (though unevenly concentrated), and it and its decay products (called radon daughters) account for about two-thirds of our annual background radiation dose. Radon-222 further decays into polonium-218 and bismuth-214, among other radioactive elements, both of which emit alpha particles. Because radon-222 is a gas, it can be inhaled, and the alpha particles it releases can be highly damaging to the lungs. Radon-222 gas gets trapped in unventilated basements. In areas with high concentrations of radium in the soil, radon-222 also enters the groundwater and evaporates (especially in hot water), so we inhale it when we shower. Radon-222 and related radionuclides are estimated to be the most common cause of lung cancer deaths in nonsmokers. (Some lung cancer in nonsmokers is attributed to so-called passive smoking, secondhand exposure to cigarette smoke.) The Environmental Protection Agency estimates that radon causes about 21,000 of the approximately 160,000 lung cancer deaths in the United States each year. Smoking is of course the largest cause overall, and radon gas and smoking may interact to increase lung cancer risk.

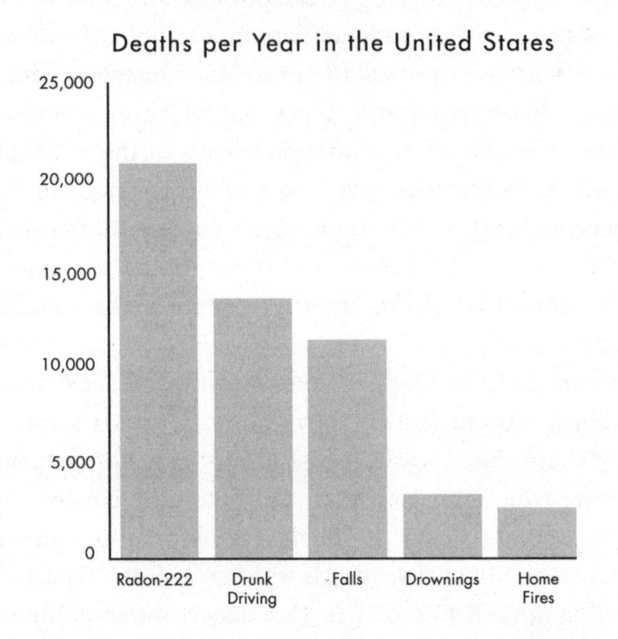

As we mentioned, the other part of the radiation we receive comes from man-made sources. About 80 percent of it comes from medical procedures like X-rays, CT scans, and nuclear medicine studies using radionuclides such as iodine-131 (thyroid scans), fluorine-18 (PET scans), technetium-99m (liver, spleen, and bone scans). (The "m" stands for metastable, which means that it decays rapidly into gamma rays.) The other roughly 20 percent of man-made radiation comes from televisions, computer screens, smoke detectors, heart pacemakers, porcelain teeth, and the like. Only a small part of our average man-made radiation exposure comes from occupational sources and the residual fallout from atmospheric nuclear weapons tests. About 1 percent of the 20 percent is the result of the nuclear fuel cycle—mining, transport, fissioning, and waste of the nuclear fuels that run nuclear power facilities.

Scientists know that the human body can tolerate a considerable amount of ionizing radiation, because we absorb a great deal that is naturally present in our environment. Still, it is wise to keep your exposure to radiation as low as is reasonably possible.

We aim to examine some of the cancers that ionizing radiations can cause—most commonly but not exclusively cancers of the blood (leukemia), breast, thyroid, lung, and brain; sources of ionizing radiation include uranium-235 (used in nuclear power facilities and the source of energy for the bomb dropped on Hiroshima); plutonium-239 (which originates in nuclear reactors and is produced by the capture of extra neutrons by uranium-238 to form uranium-239, which then undergoes a series of decays to form plutonium-239; it is also used as fuel in nuclear power facilities and was the source of energy in the bomb dropped on Nagasaki); iodine-131 (a radioactive by-product of fission that is released by nuclear power plant facilities and is a cause of thyroid cancer); cesium-137 (used in radiation therapy but deadly if not properly contained); polonium-210 (one of the cancer-causing agents in cigarette smoke); and strontium-90 (another by-product of uranium-238 fission, which settles in the teeth and other bones and can cause bone and other cancers). We will look at the risks associated with other topics of broad concern: radiated food; radiation therapy given to treat cancer; nuclear power facilities; radioactive waste; and nuclear terrorism. At the end of the book, we will answer commonly asked questions.

We also will consider the risks and advantages of using nuclear energy to generate electricity. An understandable response to the accident at the Fukushima Daiichi nuclear power facility and the release of radiation into the air, ground, and sea is to question the wisdom and safety of nuclear energy in general and to ask whether we should abandon its use. In Japan, there is a powerful movement to halt the use of nuclear energy; the last of Japan's more than fifty commercial reactors, which provided 30 percent of its electricity, was shut down in May 2012. (The next month, however, two reactors were reopened.) In May 2011, the German government voted to shut all of that nation's nuclear plants by 2022, though it will continue to buy electricity produced by nuclear energy from its next-door neighbor France. Others argue that nuclear energy is essential to power the exponentially growing global need for electricity. In October 2011 the French Alternative Energies and Atomic Energy Commission (CEA) announced plans to construct a sixtieth nuclear reactor and is looking to sell its decades of technological expertise in building nuclear power plants to India, China, Britain, Poland, South Africa, Turkey, and Brazil.

The specter of radiation is so frightening to many people that it eclipses reality. We tend to forget that the 9.0-magnitude earthquake in Japan that triggered the failure at the Fukushima Daiichi nuclear power facility was so unimaginably powerful that it moved the entire 137,893 square miles of the main island of Honshu eight feet eastward; that the tsunami obliterated whole towns, drowned almost 20,000 people, and left more than 300,000 temporarily or permanently homeless; and that although additional radioactive leaks from the crippled nuclear reactors still pose a danger, no one has died from radiation poisoning, and the predictions are that the released radiation may cause only a slight, probably undetectable, increase in cancer risk in the exposed population over the next several decades. The stakes for not making intelligent decisions about radiation are so high that we must equip ourselves with as much knowledge as we can and not let our sometimes illogical fears influence our judgment or allow us to fall victim to the social, economic, and psychological damage they can inflict. But we must also carefully weigh the risks of nuclear technologies and seek the best balance of benefit and risk.

CHAPTER 1

ASSESSING THE RISKS

HOW CAN I DETERMINE MY RISK OF CANCER FROM RADIATION, AND WHY IS THERE SO MUCH DISAGREEMENT AMONG EXPERTS?

On July 16, 1945, in the Jornada del Muerto (Journey of Death) desert near Alamogordo, New Mexico, the fiery explosion of the Trinity test—the first atomic bomb—generated a light brighter than any ever seen on Earth. As it dimmed, it revealed a mushroom cloud of vaporized water and debris that grew thousands of feet into the air. J. Robert Oppenheimer (1904–1967), who more than anyone else was responsible for building the weapon, wrote afterward that watching the explosion brought to mind two lines from the sacred Hindu scripture the Bhagavad Gita: "If the radiance of a thousand suns were to burst into the sky that would be like the splendor of the Mighty One." And: "I am become Death, the shatterer of worlds." (It is perhaps more likely that his first thought was, *Wow! Thank God, it worked!*)

That shattering burst of energy was also an act of creation: it produced radioactive forms of natural elements that—apart from laboratory work during the bomb's development—had never before existed on Earth, including cesium-137, iodine-131, and strontium-90. During the months that followed these newly created radionuclides circled the globe and silently entered the bodies of everyone alive. And because some of these radionuclides remain radioactive for hundreds or thousands of years, the children of these people, their children, and all humans from that date until our species ceases to exist will have radionuclides created at the Trinity explosion in their bodies. The same is true for the radionuclides released by the more than 450 atmospheric nuclear weapons tests carried out by the United States, the Soviet Union, Britain, France, and China between 1945 and 1980, and from several nuclear power facility accidents. Of course, the amounts of radionuclides released from each of these sources differ vastly. It is inappropriate to consider atomic weapons and nuclear power facility accidents comparable, because the quantity of radionuclides released varies greatly, because they are not uniformly distributed over the Earth, and because different people have different likelihoods of encountering them.

Some of the radionuclides released by nuclear weapons testing and by nuclear power facility accidents can cause cancer. But some of the same radionuclides are used to diagnose and treat cancers and save lives. What is the balance between the potential harm and benefit posed by radionuclides and by all forms of radiation?

To determine whether this balance favors harm or benefit, it is necessary to know what radiation dose a person has received. This is not as simple as it might seem (in fact it is exceedingly complex, even for radiation experts), so we ask the reader please to bear with the following several pages of technical information, knowing that in the end all you really need to remember is one technical term: millisievert (mSv), named for the Swedish medical physicist Rolf Maximilian

Sievert (1896–1966), who did pioneering work on the biological effects of radiation exposure. A sievert (Sv) is a unit of radiation. Each year we generally receive a few thousandths of a sievert, called a millisievert. People in the United States on average receive 6.2 mSv of radiation annually.

Radioactivity is measured by the number of atoms decaying (losing energy by emitting radioactive particles and/or electromagnetic waves) in a certain amount of time. The disappearance of a radionuclide is measured by how long it takes for one-half of its atoms to decay. That can take a long time, as something can be reduced by one-half almost forever, until only one atom remains—and then it decays. But most of the starting radioactivity is gone after about 10 half-lives; only about one-thousandth of the starting radioactivity remains.

These measurements have many names, depending on what you want to quantify. At first it is easy to mistake which unit to use, so one can end up comparing the radioactive equivalent of eels to elephants. It is also easy to mistake amounts: 1 microsievert (a millionth of a sievert) is a thousand times smaller than 1 millisievert (a thousand mSv make 1 Sv), yet several news reports of the Fukushima Daiichi nuclear power facility accident confused these units.

In estimating how a radiation exposure might affect us, scientists need to consider the amount of radiation we are exposed to; what type of radiation it is; how much of it gets into the various cells, tissues, and organs in our body; and how susceptible these tissues and organs are to radiation-induced damage. Some cells, like bone marrow, skin, and gastrointestinal tract cells, are especially sensitive to damage from radiation. One reason is that they divide frequently—rapidly dividing cells are more sensitive to radiation that damage DNA. For example, a normal person needs to produce about 3 billion red blood cells each day to stay healthy. Other cells, predominantly those that divide infrequently, if ever, like heart, liver, and brain cells, are relatively resistant to radiation-induced damage.

So to determine the amount of radiation in an exposure, we must calculate the quantity of radiation emitted or released from a source, be it a CT scanner, a radiation therapy machine, a nuclear weapon, a failed nuclear power facility, or a radioisotope injected for a PET scan.

But calculating the quantity of radiation is complex. Some diagnostic radiation machines emit electromagnetic waves such as X-rays or particles such as protons, neutrons, or electrons. Other radiation-related activities, like fissioning uranium-235 or plutonium-239 in a nuclear weapon, emit gamma rays and neutrons. Most fission products emit electrons and gamma rays. The explosion of the Chernobyl nuclear reactor released into the environment more than 200 radionuclides in diverse physical and chemical forms, including radioactive gases such as xenon-133 and iodine-124 and -131, as well as radioactive particles. These gases rapidly disperse into the atmosphere. The radioactive particles also disperse across a very broad area—unless it happens to rain when the radioactive cloud passes over you and particles of cesium-137 and strontium-90 fall to the ground with the raindrops.

Unfortunately, it was raining when the radioactive plume from the 1986 Chernobyl accident containing particles with iodine-131 and cesium-137 passed over Scotland. Consequently, substantial amounts of these radionuclides landed on grass. The grass was subsequently eaten by grazing animals, especially sheep, and those radionuclides were incorporated into their bodies and secreted in their milk. The iodine-131, with an 8-day half-life, was gone in about three months. But the cesium-137 was concentrated in the meat of the sheep, and with its half-life of 30 years, it stayed around for the lifetime of the sheep. The level of cesium-137 in many of these animals exceeded government safety standards; consequently many sheep were killed and buried, and their meat was quarantined from human consumption.

For a radioactive substance or radionuclide, like a gram of radium-226 or a gram of cesium-137, we can compute how much radiation it releases by considering the number of spontaneous disintegrations (decays) that occur in the nuclei of the atoms in that gram over a certain time interval, for example, one second. This rate of decay, referred to as the amount of radioactivity in radionuclides, is measured in units called *becquerels* (Bq), named after the nineteenth-century French physicist Antoine-Henri Becquerel (1852–1908). One becquerel equals one nuclear disintegration per second. Because this is an extremely small quantity, scientists often speak of thousands of becquerels (a kilobecquerel, KBq), millions of becquerels (a megabecquerel, MBq), a million million becquerels (a terabecquerel, TBq), or even a billion billion becquerels (a exabecquerel, EBq). It's like an expression of speed. If a becquerel is a person walking 1 mile per hour, a kilobecquerel is like the same person walking (or rocketing) 1,000 miles per hour, and so forth. However, the quantity of becquerels a substance contains is not the only consideration for human health. Because different nuclear disintegrations release different electromagnetic waves and particles, the same quantity of becquerels released can have substantially different potential health consequences. Also, not all radioactive substances are equally radioactive. When we compare similar quantities of thorium-230 and uranium-234, for example, the thorium-230 is about 1 million times more radioactive—it has 1 million times more disintegrations per second.

Once we have ascertained the amount of radioactivity, we must determine how much radioactive energy it deposits into something. That "something" can be the air, another substance, or our bodies.

Then we must determine how this radioactivity interacts with humans. This is referred to as *dose,* which is quite different from *emitted radiation.* Imagine a gram of cesium-137 inside a lead box. It is releasing radiation in the form of electrons and gamma rays, but no one is being exposed to it, because these radiations cannot penetrate the lead. So the dose of radiation to any person is zero, and hence it has no chance for harm to us. But if you are holding this same gram of cesium-137 in your hand, the same electrons and gamma rays that it emits through the spontaneous decay of its nucleus will interact with the skin, muscle, and nerve cells in that hand. And because gamma rays can travel considerable distances and pass through many substances, other parts of your body will be exposed to radiation, although not uniformly. As these gamma rays pass through your cells, they will deposit some of their energy within each cell they strike. This amount of energy is the radiation dose to the cell.

Another concept in radiation dosimetry is the *radiation absorbed dose,* which is expressed in units of grays (Gy), after the British physicist Louis Harold Gray (1905–1965). A gray is the amount of energy a dose of radiation deposits in a tissue. We will skip over them other than to say that the quantity of grays absorbed into a tissue or organ (adjusted for some biological factors) can be converted to a number of sieverts, the unit used to estimate risk of harm, like cancer, from radiation exposure.

Finally, determining the *effective dose,* measured in sieverts, considers two issues. First, not all types of radiation are equally damaging—for example, a dose of neutrons absorbed is much more damaging than the same dose of X-rays. Second, different cells, tissues, and organs in the body, as we saw, have different sensitivities to radiation damage. *Effective dose* adjusts for these variables and thereby gives a better estimate of the potential harmful consequences of a radiation exposure. This, at last, concludes our discussion of units of radioactivity. But please try to remember millisieverts, as we will translate everything into them from now on.

Having slogged through so many technicalities, let's examine how scientists analyze radiation emitted, energy absorbed, and biological damage from that radioactivity, so that we can make the one judgment that really matters: What am I exposed to, and is it bad for me? We'll use a basketball analogy, for simplicity.

When a player sends a basketball through the hoop, the number of points awarded can vary. A free throw is worth 1 point, a basket shot from inside a circumscribed area is worth 2 points, and a basket shot from outside that area is worth 3. The team's score in a game is not the number of times its players sent the ball through the hoop but the total of the points awarded for those baskets. It's the same with measuring radioactivity: the amount you are exposed to is not necessarily the amount that you will absorb, and that is not necessarily directly correlated with the amount of harm. Determining that final harm score means weighing and balancing several factors.

If there were a direct correlation between a specific amount of exposure and the onset of disease, a simple chart would clarify things for you. But the relationship between radiation and disease is not so simple.

The conventional approach to determining a person's cancer risk from a radiation exposure is to compare the range of possible effects from the dose in the scientifically accurate but difficult-to-understand units we've detailed. When there is a nuclear or radiation accident, public health authorities often give information in terms of what radiation dose people received (or will receive in the future) and/or how much radioactivity is in something they may encounter, such as food or water. They then compare these doses or amounts of radioactivity to a benchmark, such as the normal background radiation dose, or the dose a nuclear power facility worker receives annually, or the regulatory limit or threshold for radioactivity in food or water.

Such information, given to people who are not radiation scientists or physicians, is likely to be uninformative at best and misleading at worst, and it is at once confusing and simplistic. The implication is that if you receive a dose similar to or less than your normal background dose, or less than a regulatory limit for food or water, you need not worry. For example, if the regulatory limit for radioactivity in milk is 500 Bq per liter and the milk you are drinking contains 350 Bq per liter, you are not at risk.

But things are not so simple. For any radiation dose, the risk of getting cancer also depends on one's age at the time of exposure, estimated remaining life span, exposure to other cancer-causing agents (like cigarette smoke), concurrent health problems that can be exacerbated by radiation, and other complicated variables. Simply put, the implications for an eighty-year-old exposed to a given dose of radiation are entirely different from those for a three-year-old who receives exactly the same dose.

Assessing risk requires statistical analyses. You cannot rely only on dose to express a person's risk of getting cancer, because dose is only an intermediate quantity between their radiation exposure and their cancer risk. A more helpful way to link cancer risk to exposure is to specify a person's lifetime risk of cancer regardless of the cause; specify the additional lifetime risk resulting only from a specific radiation exposure; estimate future cancer risk for persons exposed in the past (or who soon will be exposed) and who are currently free of cancer, radiation related or not; or estimate the likely increase in numbers of cancers in an exposed population such as people evacuated from Fukushima.

When we talk about the dangers of radiation, we are usually referring to ionizing radiations, which can alter the structure of atoms, molecules, and chemicals in our cells and cause cancers. Most data suggest that exposure to nonionizing radiations (except UV), like those from TVs, computer screens, high-voltage electrical transmission wires, and the like, are not harmful. This area is controversial and conclusions may change, but the adverse effects of nonionizing radiations, if any, are unquestionably small compared to the proven harmful effects of ionizing radiations such as neutrons and gamma rays. The challenge in considering risk of illness from a new exposure to an ionizing radiation—say, from a radiation accident—is to compare it to voluntary and involuntary cancer and noncancer risks in everyday life, like driving a car, riding a motorcycle, flying in a jet aircraft, or going into a basement containing radon gas. By looking at the whole picture, we can weigh the cancer risk from a radiation exposure and decide whether a past exposure is important or whether a future exposure is acceptable.

Equally important, we must compare the risk of cancer (and the uncertainty surrounding it) with potential alternatives and potential benefits. For instance, someone who has a CT angiogram scan of his heart is exposed to about one-tenth the amount of radiation as the average survivor of the Hiroshima and Nagasaki A-bombs. For someone at risk of a heart attack and sudden death, this level may be acceptable, especially if a medical intervention based on results of the scan can substantially reduce the chance of sudden death. But someone who is in no immediate danger or who has no effective intervention and just hopes for reassuring information might decide against the test. A person having six PET scans to look for cancer receives about the same dose as an atomic bomb survivor. Some people undergo several PET scans for different reasons (see chapter 6).

The opinions of scientists and physicians regarding consequences of radiation exposures, such as after a nuclear power facility accident, sometimes seem polarized, and it may be difficult for people to know which viewpoint, if any, is correct. For example, some experts may suggest that hundreds, thousands, or even hundreds of thousands of cancers, birth defects, and genetic abnormalities may develop as the result of radiation over several decades post-accident, whereas others estimate that few, if any, will result. But if we exclude experts who take extreme positions (of whom there are many and who seem to be the focus of media attention), we find that knowledgeable scientists agree more than is immediately apparent.

The sources of uncertainty in estimating consequences of radiation exposure are many, but we will highlight only a few. First, is there a threshold or trigger point, beyond which radiation exposure can increase the risk of developing cancer? Scientists agree that above a certain dose (usually about 50 or 100 mSv) there is a linear relationship between radiation dose and cancer risk: the higher the dose, the greater the risk. But they disagree heatedly over whether there is an increased cancer risk from lower radiation doses, for several reasons. For one, the increased risk, if any, from these low doses may be so small that it would take studies of millions or even billions of people to be certain such a risk exists. On the other hand, data from the atomic bomb survivors, nuclear industry workers, X-ray technicians, children receiving CT scans, and children exposed to background gamma radiations (and perhaps radon) are consistent with an increased cancer risk even at the lowest doses received. Even if many other epidemiological studies show no increased cancer risk, we always need to remember that the inability to detect an increased cancer risk in a population exposed to radiation, even a large population, is not proof there is no risk.

Despite this uncertainty, scientists and regulatory agencies generally agree to assume that even a low dose of radiation is potentially harmful and that voluntary radiation exposures should be considered in the context of a potential benefit and possible risk. This linear relationship between any radiation exposure and risk is referred to as the *linear, no-threshold radiation-dose hypothesis.*

Opponents of this hypothesis typically cite the fact that very large studies of nuclear industry workers, radiologists, and populations living near nuclear facilities have, in general, failed to show a convincing increase in cancers or other adverse health effects. The exceptions, which we just discussed, inevitably engender the greatest media attention, sort of man-bites-dog versus dog-bites-man. As evidence of this ongoing controversy, the National Academy of Sciences has recommended to the Nuclear Regulatory Commission that it might undertake a more modern and sophisticated study because of concerns of technical and/or statistical flaws in earlier studies. Pilot studies have been proposed, but whether they will be carried out or, if they are, will result in a large-scale reexamination of this question is unknown.

Opponents of the linear no-threshold radiation-dose hypothesis also argue that globally people are exposed to vastly different doses of background radiation, sometimes a tenfold difference, with no detectable difference in cancer rates. For example, residents of Ramsar, Iran, live near hot springs containing high levels of radium and radon gas; they receive more than 40 times as much background radiation annually as someone living in New York or London, yet they have no special health problems. But we lack specific estimated radiation doses for most people living in Ramsar, and additional unaccounted-for confounding variables may be present.

Some data, somewhat controversial, even suggest exposure to low doses of radiation may have health benefits. This idea is known as *hormesis.* But very few scientific data support hormesis, and most scientists are unconvinced that such a benefit exists. However, radiation exposure may have complex and competing effects on cancer risk. For example, although radiation-induced mutations can cause cancer, radiation may decrease cancer risk by killing other potential cancer cells. Also, radiation may affect the immune system and thereby increase or decrease cancer risk. We cannot discern these individual and competing effects in humans and are consequently left to determine the net effect by comparing cancer incidences in persons exposed or not exposed to radiation. Confounding this issue even further is the observation that people with different genetic backgrounds may have different cancer risks from the same radiation dose.

A second source of uncertainty comes from the difficulty in comparing the consequences of a dose of radiation given over a short interval (for example, instantaneously for the atomic bomb survivors) with the same dose of radiation given over a protracted interval, perhaps days, months, years, or even decades. Most of our knowledge of the adverse health effects of radiation comes from A-bomb survivors or persons who received radiation over a relatively brief interval in a medical context, like radiation therapy for cancer, typically given over a few weeks. Many scientists argue that a radiation dose's adverse effects are much less if the dose is given over a long interval. However, others argue that the same dose given over a long interval has a similar or even a greater effect. Some recent studies suggest that nuclear workers receiving a dose of at least 200 mSv over more than a decade have an increased risk of cancer. Recent research on nuclear workers and a thorough review of publications suggest prolonged exposure to low doses of radiation may be as harmful as (or perhaps more harmful than) the same dose given over a short interval. Estimates of adverse health effects from the same dose of acute versus chronic radiation exposure cover a very broad range; risks from prolonged exposures are said to be 4 times more to 4 times less effective in harming people than an acute exposure to radiation.

A third controversial issue is whether it is scientifically valid or appropriate to extrapolate a very small per-person risk—say, 1 in 10,000, 1 in 100,000, or 1 in 1 million—to a very large population. When one does so, individual risks that some might consider trivial are found to result in estimates of thousands of excess cancers. For example, is it valid to take the very small per-exposure risk associated with background scatter radiation from an airport security scanner and multiply it by the millions of passengers screened each day to estimate a number of cancers in American and European populations? Some scientists argue it is valid and, in fact, necessary to determine risk-to-benefit ratio for screening. Others, like the Health Physics Society, warn that this practice is scientifically invalid for radiation doses below 100 mSv. Nevertheless, this calculation may not be unreasonable if the uncertainty in the risk estimate is disclosed.

Fundamental to this issue is the concept of *collective dose,* which is the individual dose summed over the entire exposed population, or the average dose in the exposed population multiplied by the total number of persons exposed. The linear no-threshold radiation-dose hypothesis assumes that the consequences of a high radiation dose given to a small number of persons and a small dose given to a large number of persons are similar. Whether this is so is not known and may never be. If we accept that background radiation is associated with an increased cancer risk (not everyone does), then any additional radiation exposure is very likely to be associated with a further increase in cancer risk.

Most of us would like a precise estimate of our cancer risk from a radiation exposure, but the limitations of data, statistics, and our present state of knowledge do not lend themselves to precise estimates. The best we can do is estimate a range that most likely includes the correct number: say, between 500 and 2,000 extra cancers per 10 million people exposed to a range of radiation doses. For example, researchers estimate that of the approximately 170 million Americans who were alive during the Nevada atmospheric nuclear tests carried out between 1952 and 1957, between 11,000 and 270,000 developed extra thyroid cancers (mostly nonfatal) as a consequence of exposure to iodine-131. This is an important increase in thyroid cancers, but it is only a small proportion of the more than 2 million thyroid cancers diagnosed since 1952.

In short, most scientists and scientific organizations avoid (or should avoid) estimating precise numbers of events, like cancers, from radiation exposures. More often they suggest a possible range for these events. Sometimes these ranges are very large, perhaps ten- or hundredfold differences (for example, 10 to 1,000 cancers). Although people may wonder how the estimated ranges can be so large, these estimates reflect uncertainties in radiation dose, distribution, and potential biological consequences. The lower and upper ends of these estimates (10 to 1,000 cancers) show that experts disagree far less (if at all) than is sometimes emphasized in the media. Scientists can only weigh the evidence they have and reach the best possible conclusion, acknowledging that the whole truth is for now unknown and may even be unknowable.

With this in mind, we intend to set aside fears, some of which are baseless, and describe the effects of ionizing radiations by using the best data we have. Although the knowledge of radiobiology is not complete, the effects of radiation on human health have been studied extensively for decades, and we may know more about it than we know about the effects of most, if not all, other chemicals and toxins we are exposed to.

CHAPTER 2

RADIATION FROM DISCOVERY TO TODAY

A BRIEF HISTORY OF THE DISCOVERY OF RADIATION

In the late nineteenth century, radiation was an exciting curiosity; in the twentieth, scientists determined how to use it both as a boon and as a terror to mankind. In the past sixty years, our dose from man-made radiation has come to equal the dose we receive from natural background sources; medical diagnostic procedures and nuclear medicine now account for nearly half the dose that the average American receives each year. This figure is more than double what it was twenty years ago, and it continues to grow.

The nineteenth-century pioneers of radiation science were at first unaware of radiation's inherent danger or benefit and instead were enthralled by its mysterious and seemingly magical effects. In 1879 Sir William Crookes (1832–1919), an Englishman, discovered what he called "radiant matter" in a vacuum-sealed electrified glass tube of his design, which was named for him.

Wilhelm Röntgen discovered X-rays during an experiment with a Crookes tube in November 1895. He had painted a screen with bits of fluorescent crystals for use in his next experiment; when he passed electricity through the tube, the rays it emitted struck the crystals and lit them up. The crystals glowed even after he covered the tube with black cardboard to keep out all visible and ultraviolet light. Röntgen knew that cathode rays travel only about 3 inches in the air, which meant that the fluorescence so far away could not have been caused by them, at least directly. These mysterious rays traveled through any paper and wood that he placed between the source and the screen; he put a variety of objects in front of the tube to see if there was anything they did not penetrate. On one occasion, he noticed the outline of the bones in his hand shining on the wall. He was so eager to study this curiosity that he moved into his lab, eating and sleeping there so he could work uninterrupted.

Uncertain what these powerful rays were, he decided to temporarily call them X-rays, using the mathematical symbol *X* for something unknown, hoping to come up with a more descriptive name once he learned more. He discovered that X-rays are created in a vacuum cathode tube when the electrical arc created by the current passed through the tube interacts with the inert gas inside it. An X-ray image is the result of the rays' striking a solid object such as a bone; it shows the densest material the lightest, which is why the shadows of the bones in an X-ray are lighter than the penumbra of the less dense flesh that surround them. (In 1901 Röntgen received the first Nobel Prize in Physics for this discovery.)

About two weeks after his initial discovery, Röntgen put the left hand of his wife, Anna Bertha, on a photographic plate and made the first X-ray picture, or röntgengram, complete with wedding ring. Frau Röntgen was less impressed by the science than she was terrified by the sight of her bones. "I have seen my death!" she reportedly exclaimed.

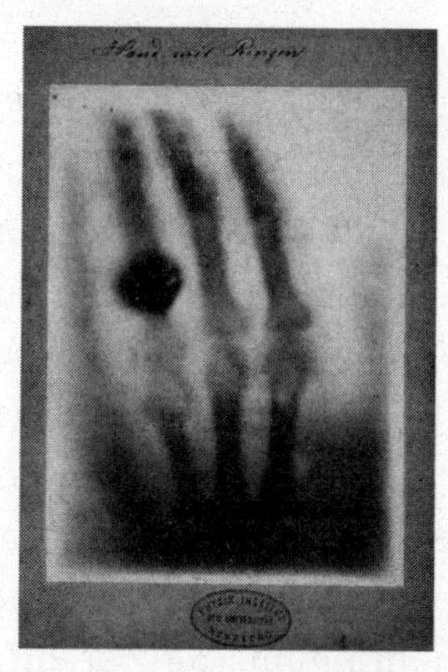

Anna Bertha Röntgen's left hand and wedding ring, considered the first X-ray photograph. (The bones at the bottom are within the palm and back of the hand.)

In December 1895, Röntgen published a paper entitled "On a New Kind of Rays: A Preliminary Communication." Other scientists quickly joined in the research into these mystical rays (at first also called Röntgen rays). In early 1896 Henri Becquerel, drawing on Röntgen's work, accidentally discovered radioactivity (but did not name it), when in preparation for an experiment that required bright sunlight, he wrapped crystals of another fluorescent substance in photographic plates that he covered with thick black cardboard to protect the plates from light. But before a day sunny enough for him to perform the experiment came along, Becquerel noticed that the photographic plates already showed the image of the crystals. This eventually led to the realization that radiation can be spontaneously emitted, the characteristic now known as "radioactivity."

Becquerel was the latest in a line of distinguished scientists. His grandfather Antoine César made advances in the fields of electricity and especially electrochemistry. His father, Alexandre, investigated solar radiation and phosphorescence. Henri followed in his father's exploration of phosphorescence and the absorption of light by crystals. Less than three months after Röntgen's discovery, Henri looked for a connection between X-rays and naturally occurring phosphorescence. Becquerel put uranium salts, which glow (phosphoresce) on exposure to light, near a photographic plate covered with light-opaque paper. (In chemistry, a salt is a compound that has been made electrically neutral.) The plate became fogged, which meant that something other than visible light had passed through the paper. After numerous experiments with uranium salts of differing composition produced the same reaction, Becquerel concluded this effect was a property of rays emanating from the uranium atoms. These rays were quickly named Becquerels. He later showed that these uranium rays differed from X-rays because exposure to an electric or magnetic field could deflect them, and because they gave an electrical charge to gases—they ionized them.

It was Marie Sklodowska Curie (1867–1934) who named radioactivity. In September 1897 she was looking for an idea for her Ph.D. thesis, and her husband, Pierre, the head of the laboratory at the Municipal School of Industrial Physics and Chemistry in Paris, suggested she investigate the curiosity described by Becquerel. Her experiments showed that the total radioactivity of some uranium and thorium salts was greater than the radioactivity from the uranium. This meant that there had to be something else in the uranium salts that emitted stronger radiations. Pierre and Marie then extracted from uranium salts tiny amounts of an until-then-unknown substance she called "radium" (after the Latin word for ray), which is more than a million times more radioactive than the same mass of uranium. She also discovered an element that decays from radium, which she called polonium, after her native Poland. Radium and polonium are the source of most of the radioactivity in uranium ore.

Madame Curie studied the radioactivity of all compounds containing the known radioactive elements, including thorium (named for the Norse god Thor), which she later discovered was also radioactive. She noticed that the strength of radiation from uranium could be measured exactly, and that no matter what compound it was in, the intensity of the radiation was proportional to the amount of uranium or thorium in that compound. This led her to the revolutionary realization that the ability of a substance to emit radiation does not depend on the arrangement of the atoms in a molecule. In 1903, for their discovery of natural radioactivity, the Curies shared the Nobel Prize in Physics with Becquerel. In 1911 Marie Curie received a second Nobel, this time in chemistry, for her isolation of radium and polonium, and for her inquiry into their chemical properties. (In 1935 her daughter Irène Joliot-Curie and Irene's husband, Frédéric Joliot, shared the Nobel Prize in Chemistry for their discovery that radiation can be induced in certain elements.)

But these important discoveries were not restricted to the Curies or to France. In 1899 and 1900, while studying radium, the New Zealander Ernest Rutherford (1871–1937), working at McGill University in Montreal, discovered alpha and beta particles. At the same time, the French physicist Paul Villard (1860–1934) discovered another form of rays that he termed "gamma rays." In 1914 Rutherford would prove that gamma rays were a form of light similar to X-rays but with a far shorter wavelength and thus penetrated deeper than the other rays or particles.

Also in 1900, British scientist Frederick Soddy (1877–1956) observed that radioactive elements spontaneously disintegrate into variants of the original. He called them "isotopes," from the Greek *iso* (equal) and *topos* (place). He also discovered that radioactive elements have what he called a half-life—and he did the first work on calculating the energy released during this decay.

At first the dangers of radiation were not apparent. In the early twentieth century, watches had radium dials that glowed in the dark. Tiny green dots of radium were painted on with fine brushes by young girls, whose small hands made them the most dexterous to do the work. These "Radium Girls" sat at a bench with a little pot of liquid radium into which they dipped their brushes. To keep the dots small and perfectly rounded, the girls licked the brush tip to get a fine point. The body recognizes radium as calcium and deposits it in bone, so many of the girls soon developed cancer of the jaw—the nearest bone. (Because radium is absorbed by bone, it can also harm bone marrow, causing severe anemia.) Many of the young women were disfigured, and some died. When five dying Radium Girls filed a suit against their employer, the United States Radium Corporation responded by accusing them of having syphilis. But as the case unfolded, it became clear that the employers had long taken precautions to protect themselves against any radiation and had done all they could to cover up the danger, even telling the workers that it was safe to use their tongue to make the points on the brushes. The case led to establishing the right of employees to sue their employer for occupational diseases. Radium, applied in a safer manner, was used until the 1960s.

Marie Curie, who constantly handled radioactive materials unprotected and worked as a radiologist using unshielded early radiology devices, died in 1934, probably of radiation-induced aplastic anemia, or bone marrow failure, a condition caused when the bone marrow does not produce enough new blood cells to replace those that have lived out their normal life span. Today her laboratory papers and cookbook still have high levels of radioactivity and are kept in lead-lined boxes. It is not known if Pierre, who also worked with radioactive material, suffered from the effects of radiation, as he died in an accident in Paris in 1906 when he slipped and fell under a horse-drawn cart.

Others have been clearly damaged by radiation. In 1986 twenty-nine nuclear facility workers and firefighters died as a result of entering the Chernobyl reactor complex. They perished within a month not only from burns from the flames but also from radiation-induced bone-marrow-failure anemia from an exponentially higher dose than Marie Curie's. (Two others were killed immediately by the explosion; their bodies were not recovered.) In addition to bone marrow failure, the firefighters and workers also suffered severe radiation damage to their lungs, gastrointestinal tracts, and skin. They also had injuries from thermal burns and trauma from the explosion.

But even with warnings like Marie Curie's death and the aftereffects of atomic weapons, radiation was often treated as a harmless curiosity. Beginning in the 1920s, for example, shoe stores had X-ray fluoroscope machines (sometimes with the trade name Pedoscope) to determine the right fit. They were sources of amusement for customers, and all family members with them, who loved to see the bones of their toes inside the outline of their shoes. Typical exposure was about 15 seconds. On average that meant about 0.5 mSv, or one-sixth the average annual dose from background radiation. The inherent danger of the machines was understood in 1949, after the birth of atomic weapons, and most fluoroscopes were phased out in the 1950s.

The years immediately following the Curies' breakthroughs were spent trying to unravel their implications, and the work brought discovery after discovery. In 1901 Soddy and Rutherford discerned that radioactive thorium was converting itself into radium. In 1904 Rutherford—the Magellan of nuclear science, discovering the bits and pieces of the orderly universe of the atom—found that alpha radiation is actually a heavy, positively charged particle with two protons and two neutrons.

The decade between 1905 and 1915 brought important advances in understanding the nature of atoms and subatomic particles. Robert Millikan (1868–1953) showed how to measure the electric charge and mass of an electron. Rutherford developed his theory of the structure of atoms. Soddy and the Polish-American Kasimir Fajans (1887–1975) separately developed the theory of isotopes of elements, and Fajans explained radioactive decay. (Seven years later, in 1919, Francis Aston [1877–1945] proved the existence of isotopes.) In 1914, at the start of World War I, H. G. Wells's novel *The World Set Free* imagined an atomic war in 1956 that destroys the major cities of the world.

In 1919 Rutherford accomplished the first artificial nuclear reaction, which happens when particles from the radioactive decay of one element are used to transform the atomic nucleus of another, a process known as *transmutation.* He was hailed as having "split the atom," although his work was far short of the nuclear fission reaction that comes from uranium and other "heavy" elements (those with properties of metals). For all his brilliance, however, Rutherford did not believe transmutation of atoms could be a source of power. In 1933 Leó Szilárd (1898–1964), a Hungarian-born physicist who had fled Nazi Germany and taken refuge in London, was the first to theorize that "if we could find an element, which is split by neutrons, and which would emit two neutrons when it absorbs one neutron, such an element, if assembled in sufficiently large mass, could sustain a nuclear chain reaction." He did not envision nuclear fission as one of these neutron-producing reactions, since this reaction was not known at the time.

Szilárd filed to patent "the liberation of nuclear energy for power production and other purposes through nuclear 'transmutation'" in 1934. He amended his patent application the next year, adding that uranium and bromine are "examples for elements from which neutrons can liberate multiple neutrons." He hoped to keep the contents of the patents secret, but when he learned that to guarantee it he would have to assign the patents to an agency of the British government, he offered them to the War Office. His offer was refused because, he was told, "There appears to be no reason to keep the specification secret so far as the War Department is concerned." A few months later the British Admiralty wisely accepted the patents.

In 1938 Lise Meitner (1878–1968), her nephew Otto Frisch, and others discovered that uranium can capture neutrons, then form unstable products and undergo fission, which causes the ejection of more neutrons in a continuous chain reaction. Meitner, an Austrian Jew who worked in Germany and then fled to Stockholm to evade the Nazis, is often thought to have been denied, because of her religion, part of the 1945 Nobel Prize in Chemistry awarded to the German Otto Hahn (1879–1968).

There are two kinds of chain reaction. The first is fission, the source of atomic bombs and nuclear power, in which matter is converted into energy. The nucleus of an atom, struck by a neutron and absorbed, splits into two pieces, one lighter than the other, producing gamma rays and releasing a tremendous amount of kinetic energy. Other neutrons, slowed from their natural pace to travel at just the right speed, strike other nucleii and keep the reaction going.

The second kind of energy in a chain reaction is called

fusion: the joining of two or more things to create a single thing. In nuclear physics, fusion is a chain reaction in which the nuclei of two or more elements fuse to form a single nucleus of a new element and simultaneously release energy. This happens every instant within the Sun, and it happens in a hydrogen bomb. The difference between fission and fusion is that fusion requires a great deal more energy to start the chain reaction, but fusion also yields vastly more energy—a hydrogen (fusion) bomb is roughly one thousand times more powerful than an atomic (fission) bomb.

The discovery of fission was announced in January 1939 (though the discovery was in 1938). As soon as J. Robert Oppenheimer heard about it, he understood that an atomic bomb was possible. He was neither the only one nor the first. Szilárd immediately realized the possibility of using neutron-induced fission to sustain a chain reaction. In 1939 Szilárd and Italian-born Enrico Fermi, who had emigrated to the United States to protect his Jewish wife from Italian Fascists, proved this concept using uranium. In this reaction, a neutron plus a fissionable atom (uranium-235) causes a fission resulting in more than one neutron being released. These "excess" neutrons then fission other uranium-235 atoms, resulting in a nuclear chain reaction. Under appropriate circumstances, as in a nuclear power facility, the speed and extent of this reaction can be controlled by regulating the density and speed of the neutrons released and the concentration of the uranium-235 fuel. However, in some circumstances this reaction is self-propagating and thus self-sustaining. This is the principle when a nuclear reactor core melts down or an atomic bomb explodes.

The first man-made demonstration of a self-sustaining nuclear chain reaction was accomplished by Fermi and others in a laboratory beneath the football stadium at the University of Chicago in late 1942. The initial work on building an atomic bomb was carried out at Columbia University and in other parts of Manhattan, before the large team was assembled at Los Alamos, which is why the top-secret enterprise is known as the Manhattan Project.

With the announcement of the discovery of fission, Szilárd saw the urgent need for atomic bomb research. He hoped that Fermi, who won the 1938 Nobel Prize in Physics for his "demonstrations of the existence of new radioactive elements produced by neutron irradiation, and for his related discovery of nuclear reactions brought about by slow neutrons," would write to President Franklin Roosevelt to explain the opportunity at hand. Fermi was reluctant, fearing it would jeopardize his and his wife's émigré status. Szilárd did not feel that he had the stature to write to Roosevelt, so he enlisted Albert Einstein. Einstein's letter of August 1939—written with Szilárd's help—set the United States on the road to developing nuclear weapons.

THE A-BOMBS

On August 6, 1945, six and a half years after the announcement of the discovery of fission and just two weeks after the Trinity test in New Mexico, "Little Boy" was exploded over Hiroshima. Three days after that, the bomb named "Fat Man" was dropped over Nagasaki. (The bombs were set to explode about one-third of a mile above the cities, to give them added force from the blasts' downward effects.) 150,000 to 240,000 people were killed, half of them during the first day. It is widely believed that radiation caused most of these deaths, but this is incorrect.

Albert Einstein
Old Grove Rd.
Nassau Point
Peconic, Long Island

August 2nd, 1939

F.D. Roosevelt,
President of the United States,
White House
Washington, D.C.

Sir:

Some recent work by E.Fermi and L. Szilard, which has been communicated to me in manuscript, leads me to expect that the element uranium may be turned into a new and important source of energy in the immediate future. Certain aspects of the situation which has arisen seem to call for watchfulness and, if necessary, quick action on the part of the Administration. I believe therefore that it is my duty to bring to your attention the following facts and recommendations:

In the course of the last four months it has been made probable - through the work of Joliot in France as well as Fermi and Szilard in America - that it may become possible to set up a nuclear chain reaction in a large mass of uranium,by which vast amounts of power and large quantities of new radium-like elements would be generated. Now it appears almost certain that this could be achieved in the immediate future.

This new phenomenon would also lead to the construction of bombs, and it is conceivable - though much less certain - that extremely powerful bombs of a new type may thus be constructed. A single bomb of this type, carried by boat and exploded in a port, might very well destroy the whole port together with some of the surrounding territory. However, such bombs might very well prove to be too heavy for transportation by air.

The United States has only very poor ores of uranium in moderate quantities. There is some good ore in Canada and the former Czechoslovakia, while the most important source of uranium is Belgian Congo.

In view of this situation you may think it desirable to have some permanent contact maintained between the Administration and the group of physicists working on chain reactions in America. One possible way of achieving this might be for you to entrust with this task a person who has your confidence and who could perhaps serve in an inofficial capacity. His task might comprise the following:

a) to approach Government Departments, keep them informed of the further development, and put forward recommendations for Government action, giving particular attention to the problem of securing a supply of uranium ore for the United States;

b) to speed up the experimental work, which is at present being carried on within the limits of the budgets of University laboratories, by providing funds, if such funds be required, through his contacts with private persons who are willing to make contributions for this cause, and perhaps also by obtaining the co-operation of industrial laboratories which have the necessary equipment.

I understand that Germany has actually stopped the sale of uranium from the Czechoslovakian mines which she has taken over. That she should have taken such early action might perhaps be understood on the ground that the son of the German Under-Secretary of State, von Weizsäcker, is attached to the Kaiser-Wilhelm-Institut in Berlin where some of the American work on uranium is now being repeated.

Yours very truly,

A. Einstein

(Albert Einstein)

Although certainly those people with the most severe radiation exposures died shortly afterward, radiation released by the bombs was not the cause of most deaths. The immediate effect of a fission bomb is superfires from the intense heat generated and the potent concussive wave that moves away from a bomb—the same effects, though vastly greater, of a conventional bomb. About 60 percent of the immediate deaths (and about 90 percent overall) in Japan resulted from the force and fire, and the majority of the remainder from falling debris. About 10 percent of deaths are attributed to radiation.

Masao Tomonaga, a Tokyo physician friend of Bob and a world-famous hematologist, was a two-year-old living in Nagasaki the day the bomb was dropped. His father, a physician in the Japanese Army Air Force, was in Taiwan. Masao and his mother lived in a typical Japanese wooden house, a mile and a half from ground zero.

"A small hill behind the house protected us against the terrible blast wind," he wrote to us in April 2012,

> but part of the house was knocked down and according to my mother's memory, within ten minutes the house caught fire, and we escaped to a nearby shrine. Unfortunately (?) [the question mark is his], I have no memory of the bomb's effects. It totally devastated Nagasaki Medical College, 600 meters from the epicenter, and about 900 professors, students, and nurses were killed. There are thousands of records written by survivors about their direct experience with the heat and blast generated by the bomb; descriptions of skin flash burn and damaged skin coming off in pieces are the most frequent. The wind blast caused instantaneous death and severe bodily injuries including hundreds of skin cut-wounds due to flying broken glass.

Horrific as this description is, it is not that dissimilar to those of the firebombing of Dresden and Tokyo during World War II. It is Masao's description of what happened next that expresses the particular horror of an atomic weapon. Conventional bombs had been dropped on Nagasaki before the A-bomb. People knew the resulting effect on bodies. What occurred in the days following the A-bomb blasts was completely new. Perhaps it is the effects of one big bomb, rather than many small ones, with deferred damage whose cause cannot be seen that makes nuclear weapons and radiation in general all the more terrifying.

> The earliest sign of high-dose radiation was, as you know very well, hair loss that began a week or two later. Then severe diarrhea started with bloody stool due to bone marrow failure. The next step was high fever due to infection when neutrophils [white blood cells that fight infection] declined severely. In the case of the Nagasaki bomb, 35,000 people died within a day and another 37,000 within three months. Almost the same number of people survived but after three years developed leukemia, and solid cancers of various organs after 15–66 years (even now).

Masao has spent his professional life studying long-term cancer consequences of the atomic bomb explosions and was head of the Atomic Bomb Disease Institute at Nagasaki University.

A survey in 1950 estimated that 160,000 people survived the Hiroshima bomb and 125,000 survived the one over Nagasaki. The American Medical Association estimates that more than 40 percent of the survivors were still alive in 2011, sixty-seven years later, and that 80 percent were exposed before age twenty. About 93,000 survivors of the two blasts have been—and still are—closely monitored over their lifetime. Radiation received by the survivors ranged from less than a normal medical procedure to doses large enough to cause bone marrow failure. The average dose of survivors in Hiroshima was 200 mSv or about 30 times the average American's annual radiation dose. It is also about one-half of the amount of radiation the average American receives in a lifetime, but it was received over an instant rather than over seventy-five years and adds to normal lifetime radiation dose. As many as 160 people with both extraordinarily bad *and* good luck are thought to have been in Hiroshima and in Nagasaki the day the bombs were dropped, and survived.

Leukemia was the first radiation-related cancer detected in the A-bomb survivors. Japanese physician Takuso Yamawaki in Hiroshima noted an increase of leukemia among his patients in the late 1940s. He wrote to his Western colleagues about his observations and published his experience in the medical literature. This aroused considerable interest and led to the establishment of a registry of leukemia and related disorders (originally called the Atomic Bomb Casualty Commission, or ABCC, and now called the Radiation Effects Research Foundation, RERF, funded by the Japanese and U.S. governments). Reports of increased leukemia risk were published in the early 1950s.

The ABCC set up several studies of the A-bomb survivors. To determine the extent of any increase in leukemia or other cancers, it was necessary to know how frequently these diseases occurred in a *control* population that was not exposed to radiations from the A-bombs. People normally living in Hiroshima and Nagasaki who were out of town when the A-bombs exploded where chosen as controls.

These groups—A-bomb survivors, their children, and their luckier neighbors—comprise the Life Span Study

(LSS) and have been evaluated annually for health problems including cancer, heart disease, birth defects, and heritable genetic abnormalities. Of the about 120,000 people in the LSS, about 93,000 are A-bomb survivors; the other 27,000 were in neither city when the bombs exploded and are the control group. This information is of considerable import, not just for those being studied but also because it represents much of what we know about the effects of high doses of radiation on humans delivered in a very short period of time. These data are used to develop regulatory standards and to estimate effects of nuclear and radiation accidents.

Although the risk of many cancers is increased in the A-bomb survivors, leukemias are special, for several reasons. Risks for radiation-induced leukemia differ in two major respects from those for most other radiation-related cancers. First, radiation caused a larger proportional increase in leukemia rates than in the rates of other cancers. Cells in the bone marrow are especially sensitive to cancer-causing mutations from ionizing radiation. We know that a greater proportion of the leukemia cases seen were caused by radiation because "naturally occurring" leukemia is a relatively rare cancer. Of the roughly 93,000 A-bomb survivors being followed, there were about 200 cases of leukemia. One-half of these (about 100 cases) are estimated to have been caused by radiation exposure. The 25 cases of leukemia in the approximately 700 people who received a radiation dose greater than 2,000 mSv were probably caused by the A-bombs.

A second important aspect of these leukemias is that they developed sooner than other A-bomb radiation-induced cancers, especially in children. Children who were 10 years old when exposed to the A-bombs had a threefold or greater risk of developing leukemia than someone who was 30 years old when exposed. Also noteworthy in the A-bomb survivors is that, in contrast to other cancers, the relationship between dose and cancer risk is not linear (a straight line) but a more complex mathematical function (called linear quadratic): high doses produced more than we would expect if the risk were linear. This is not so for other radiation-induced cancers, where the relationship between dose and cancer risk is linear.

Cases of A-bomb–related leukemia began appearing approximately 2 years after radiation exposure, peaked between 6 and 8 years after exposure, and returned to normal levels after 10 to 15 years. However, some recent data suggest that one form, chronic myelogenous leukemia, may have increased for many more years. Also, a close relative of leukemia termed myelodysplastic syndrome increased more slowly than the usual forms of leukemia, and its risk remains elevated in the A-bomb survivors more than 60 years later.

Another curiosity is that chronic lymphocytic leukemia (CLL), the most common leukemia in people of European descent, was not detected in the A-bomb survivors. CLL was also absent in Japanese who were not exposed to the A-bombs. This observation led to two important concepts. First, radiation exposure increases the risk of cancers that occur normally in a population but not cancers that are rare or absent. Second, CLL, in striking contrast to the other leukemias, is not a *radiogenic* cancer—one that can be caused by radiation. This second concept has been recently challenged by claims, not entirely convincing, of increased CLL in the Chernobyl-exposed population. This controversy is still being sorted out.

Because the relatively rapid onset of most leukemias after radiation exposure from the A-bombs contrasts with the onset of other types of cancer, which took decades to develop, it suggests that after a radiation accident, we can get an early readout of what is likely to happen later on by looking for leukemias in the exposed population.

Data on leukemia risk after the A-bomb exposures also tells us much about how radiation and cancer risk interact. For example, normally about 7 of every 1,000 Japanese will die of leukemia in their lifetime. However, in the A-bomb survivors leukemia deaths increased to 10 per 1,000 people. Thus, although the absolute number of extra cases is small (3 per 1,000 people), they represent a more than 40 percent increase, which to epidemiologists and statisticians is *very* large. Similarly, although leukemia accounts for only 1 percent of cancer deaths in unexposed people, it accounts for about 15 percent of cancer deaths in A-bomb survivors.

These data convey an important message for expressing the risk of a rare cancer like leukemia. Say the risk increases from 10 to 20 per 1,000 people. It is correct to say there will be 10 extra cancers in every 1,000 people or 1 case in every 100. If we consider that about 45 percent of males will get cancer in their lifetime, the number of cancers in 100 males will increase from 45 cases to 46 cases. This increase sounds small to most people. However, it is equally correct to say that the risk of the rare cancer has doubled (from 10 cases per 1,000 people to 20 cases per 1,000). A doubling in cancers sounds frightening to most people. An increase in cancer risk can sound small or large depending on how the data are presented and how they are understood.

Solid cancers—the more common cancers like breast and lung cancers—also increased in the A-bomb survivors compared to unexposed Japanese. Among the survivors who received the lowest radiation dose, only about 400 of about 5,500 cancer deaths (less than 10 percent) appear to have been caused by A-bomb radiations. Stomach cancer (a very common cancer in Japan) and lung cancer were the most frequent solid cancers among the survivors. Those who smoked were at an even greater risk of lung cancer. Liver cancer, also common in Japan, was the third most frequent cancer in A-bomb survivors. Risk of liver cancer was greater in persons who were exposed in their twenties than for those who were younger or older.

Risk of thyroid cancer in the A-bomb survivors was closely correlated with age at exposure. Most radiation-induced cases occurred in children younger than ten years of age when the A-bombs exploded. This is similar to the situation at Chernobyl, where almost all excess thyroid cancers attributed to iodine-131 occurred in children and adolescents who were younger than 20 years old when the accident occurred.

Not every type of cancer was increased in the A-bomb survivors. Why? There are several possible reasons. One is that cells of different tissues and organs may have different susceptibilities to radiation-induced damage, including the mutations in DNA that are typically the first step in cancer. (Some scientists argue that heritable changes other than in DNA, referred to as *epigenetic changes,* can also start the path to cancer, but this is controversial.) Another possibility is that mutations in DNA occur with equal frequency in cells in all tissues and organs but that some cells are better able to repair the mutations than others. Both concepts may operate.

Still another possible explanation is that A-bomb survivors may be too few in number to show a small increase of a rare cancer. Failure to find an increase of cancer in a epidemiological study does not mean there is no increase, but it does tell us that any increase must be small.

Consider, for example, bone cancers. Substantial data indicate bone cancers increase after exposure to very high doses of radiation, doses greater than 60,000 mSv, usually given to a small part of the body like a limb. However, such a high exposure given to the whole body would have resulted in immediate death in the A-bomb population, where the average dose was 200 mSv, or 300 times less. Consequently, even though bone cancers are radiogenic, we cannot know if bone cancers in the A-bomb survivors did not increase because the dose was too low to cause bone cancer, or because the number of survivors was too few to detect an increased risk, or both.

Of those 160 people to have survived both the Hiroshima and Nagasaki A-bombs, Tsutomu Yamaguchi, the last known, died in January 2010 of stomach cancer, at age 93. It is impossible to know whether his cancer was related to his A-bomb exposures.

THE EVOLUTION OF FEAR OF RADIATION

In 1954 the world's first nuclear power facility was opened in Obninsk, near Moscow; there are now over 430 worldwide. The medical use of radiation has gone from simple X-rays to CT and PET scans.

How did radiation go from being something that most people admired and wished to experience to something that most people fear and want to escape? There is, of course, no simple answer, and people's behavior is sometimes contradictory. For example, parents might worry that their child's sleeping under an electric blanket may cause leukemia from radiation, but they may also insist, often contrary to medical advice, that a head CT scan be done to exclude the possibility of a brain cancer after their child has a seizure. A head CT scan delivers a dose of radiation to the head equivalent to that of someone who was about four miles from the Hiroshima A-bomb, whereas an electric blanket emits no ionizing radiation. Similarly, people concerned about global warming are often firmly opposed to nuclear energy, yet it is the only immediately available energy source able to substantially reduce carbon dioxide emissions, albeit with some inherent but potentially solvable problems.

Several considerations may help explain this paradox. The discovery of radiation by Röntgen and others opened great horizons. Radiation allowed people to see into objects and observe their workings. Medical use of radiation could show bone and other internal structures, aid diagnoses, and save lives. Later, radiation therapy was developed, and for some cancers, it still offers the best chance of cure. Some people continue to go to radium hot springs and caves for treatment of chronic diseases such as eczema and fungal infections. In the 1940s and 1950s, physicians used radiation to cure ringworm and to shrink an enlarged thymus and tonsils (mistakenly) believed to cause recurrent upper respiratory tract infections in children.

It is difficult to say precisely when enthusiasm for radiation began to change. Even at the outset, there was evidence of potentially harmful effects of too much radiation, such as the death of Marie Curie from aplastic anemia and cancers in some of the early radiologists and in the radium dial painters. Certainly a major shift occurred after the atomic bombs were exploded in Japan. During the 1930s, when the potential of developing an atomic weapon became apparent, several physicists voiced concern about the morality of this enterprise, including Szilárd, the German Werner Heisenberg, the Dane Niels Bohr, and Einstein, among others. However, the threat that Germany would develop an atomic bomb before the United States did, the surprise Japanese attack on Pearl

Harbor, and the huge loss of lives in the Pacific War overrode these reservations.

When an army or government undertakes a strategic project, the project often takes on a life of its own that eclipses the will of the participants. Once the building of the A-bombs started, there was probably no way to stop their being detonated over Japan short of an unconditional Japanese surrender. (This unstoppable force is nicely portrayed in the 1989 movie *Fat Man and Little Boy,* directed by Roland Joffé.) And most Americans were happy to see the Pacific War end quickly; only years later did they have second thoughts about civilian casualties from the A-bombs.

The shift in the view of radiation from mankind's helper to menace was accelerated by the unfortunate lack of trust between the United States and the Soviet Union immediately after the A-bomb explosions. It is impossible to know if the nuclear arms race would have developed had the Soviets been allowed access to the Manhattan Project or had President Harry Truman and Prime Minister Winston Churchill agreed to share nuclear technologies with the Soviets. But once one country had the bomb, human nature dictates that others would want it as well. What is certain is that the secrecy surrounding the development of nuclear weapons, and as well as nuclear warships and submarines, led to growing distrust between the United States and Britain, on the one hand, and the Soviet Union, on the other; to nuclear weapons escalation; and to a rising and persistent public distrust of their government's nuclear weapons policies, especially atmospheric testing. This distrust unavoidably affected public opinion on nuclear energy policy.

Because of the way governments and industry handled the development of nuclear energy, it unfortunately became confused with nuclear weapons. On December 8, 1953,

President Dwight Eisenhower gave his famous "Atoms for Peace" speech to the UN General Assembly. He discussed the peaceful uses of nuclear energy in agriculture, in medicine, and especially in the generation of electricity. He predicted that in the future electricity would be so cheaply made, it could be provided for free. Unfortunately, this has not happened. Nuclear energy meets a substantial amount of the energy needs of many developed countries, but its progress has been troubled. Many citizens perceived government regulators and industry as not transparent in safeguarding the public. Some of these concerns are warranted, whereas others are not. The nuclear facility accidents at Three Mile Island, Chernobyl, and Fukushima Daiichi further heightened global radiation fears.

Confusion between nuclear energy and nuclear weapons is even more complex. Since the early 1990s, Iran and North Korea have been accused of claiming to be developing only nuclear power facilities while actually enriching uranium or producing plutonium to make nuclear weapons. Many people see a direct connection between the supposedly legitimate use of nuclear technology and, in nations such as Iran and North Korea, the goal of building nuclear weapons. This attitude leads inevitably to another, perhaps even greater concern: a direct correlation between the diffusion of nuclear energy technologies and the ability of nations to develop nuclear weapons. This concern is real; it cannot be ignored.

Finally, nuclear energy and nuclear weapons are commonly linked to the issue of spent nuclear fuel. Concerns abound that terrorists will gain access to these materials to develop an improvised nuclear or radiological device. These concerns are furthered by discussions of developing fast breeder reactors, which produce even greater amounts of weapons-grade materials. Perhaps in part because movies and other media often portray exaggerated impacts of radiation, many people think a nuclear power facility can explode as if it were an atomic bomb, even though this is impossible. Certainly there have been explosions within nuclear power plants, but they were not nuclear, and they had nowhere near the effect of an explosion caused by nuclear fission. The Chernobyl reactor building was destroyed by a steam explosion, and part of the Fukushima reactor building was destroyed by an explosion of highly flammable hydrogen gas.

CHAPTER 3

THE NATURE OF RADIATION

As we mentioned, our lives are the sum of trillions upon trillions of chemical reactions that usually follow specific rules. Sometimes, however, something—ionizing radiation, for example—alters those chemicals, which can in turn alter the way they react with one another and change the way our bodies behave.

What is astonishing is how varied life is, considering how few chemical components are required for it. Of the 118 known elements (each of which has a unique atomic number based on the number of protons in its nucleus), 26 are in the human body. Of those, just 6—oxygen, carbon, hydrogen, nitrogen, calcium, and phosphorous—constitute 99 percent of each of us. (Oxygen alone is 65 percent of our mass.)

About 90 elements exist in nature, some in abundance (like copper and lead), others in only very minute traces (like francium). The remaining 28 usually can be made only in laboratories or by atomic fission (although there still are traces in nature of plutonium-244, its most stable isotope, meaning that it once existed naturally, and there are minuscule amounts of a few others). All elements with a higher number in the periodic table than thallium (atomic number 81) have radioactive isotopes, and all isotopes of elements from polonium (number 84) and higher are radioactive.

Elements are different from each other because their nuclei contain different numbers of protons. Every element has a nucleus of positively charged protons and electrically neutral neutrons, surrounded by a cloud of negatively charged electrons arranged in shells. An element's atomic number—where it appears on the periodic table—is determined by the number of protons in its nucleus. So hydrogen, which has 1 proton, is number 1; oxygen, which has 8, is number 8; uranium, which has 92, is number 92.

The atomic number is what defines an element. However, many elements come in slightly different forms, with the same number of protons but different numbers of neutrons, so although the atomic number is identical, the atomic weight—the sum of an element's protons and neutrons—differs. This alternative form of an element is referred to as an *isotope* or, if it's radioactive, a *radioisotope.* For example, all forms of iodine have 53 protons (which is what makes them iodine), but they have variable numbers of neutrons. Iodine-127, with 74 neutrons and hence an atomic weight of 127, is stable, whereas iodine-131, which is unstable, also has 53 protons but 78 neutrons, so its atomic weight is 131. Because iodine-131 is radioactive, it is referred to as a radioisotope of iodine.

Some isotopes are stable—they never spontaneously degenerate into another form of the same or a different element. Aluminum-27 is an example of a stable isotope. Atomic weight is measured in units called daltons, after the English chemist and physicist John Dalton (1766–1844), who was a pioneer of atomic theory and whose work on color blindness is why the disorder is sometimes called daltonism. A dalton is the approximate mass of one proton or one neutron, a tiny amount indeed; it would take 500 million, trillion, trillion daltons to make a pound.

Some elements and their isotopes are inherently unstable, and most emit negative electrons or positive ones called positrons, alpha particles, and/or gamma rays. Very few isotopes emit neutrons spontaneously. If a neutron or proton is emitted, the atomic number will change, and we have a new element. If the atomic number does not change, this instability will produce a new isotope of the same element. For example, uranium-238 can spontaneously emit an alpha particle. This means its atomic number decreases by two protons and it becomes thorium. Also, because of the loss of an alpha particle (two protons and two neutrons), the atomic weight is reduced to 234, so what we now have is thorium-234. This process will continue until a stable isotope—lead-206 in this instance—is formed (see pages 73–74).

Alchemists once dreamed of transmuting lead into gold. To our knowledge no one has succeded. However, the creation of new elements from existing elements is precisely what is achieved by bombarding naturally occurring elements with atomic or subatomic particles.

An interesting footnote to America's independence may be the consequence of trying to transmute lead to gold. King George III (1738–1820) had serious financial problems during his reign and turned to alchemists for help. They worked in a large laboratory under the palace and probably used arsenic to facilitate transmutation. The king liked to tinker in the laboratory but unfortunately suffered from a metabolic disorder called porphyria, in which an important part of hemoglobin is not made properly. Arsenic precipitates attacks of mental instability in some persons with porphyria. Such attacks may have led to his "madness" and his ill-considered decisions regarding the American colonies.

Radionuclides can harm us, help us, or both. Cesium-137 can cause kidney, liver, and bone cancers, but radiation therapists use it in devices to treat some cancers. Carbon-14 (which is produced in the atmosphere, mostly at altitudes between 30,000 and 50,000 feet, by the interaction of cosmic rays with nitrogen molecules in the air, and drifts down to the ground) is used to determine the age of prehistoric creatures and trees and is able to reveal metabolic irregularities that underlie diabetes, anemia, and gout. Iodine-131 in sufficient amounts can cause thyroid cancer but is also used to diagnose and treat thyroid cancer and other thyroid disorders.

THE MEANING OF "HALF-LIFE"

Normal iodine-127 is not radioactive and is essential to human health. A deficiency of iodine-127 can cause enlargement (a goiter) and decreased function of the thyroid gland. If this occurs in infancy, it results in severe mental retardation, a condition referred to as cretinism. Iodine deficiency affects an estimated 2 billion people worldwide and is the leading preventable cause of mental retardation. Many people's diets do not contain sufficient amounts of iodine-127, so beginning in the 1920s iodine-127 was added to table salt in the United States and other countries. Today this practice is mandated in many countries. When you buy a box of "iodized" salt, you are ensuring that you and your family ingest sufficient iodine-127 to prevent thyroid disease.

Unlike iodine-127, iodine-131 is unstable and undergoes spontaneous decay. If you have a bowl filled with iodine-131 atoms and wait long enough, virtually all of them will decay to xenon-131. However, we cannot predict when each of these identical atoms of iodine-131 will decay; the process is random (called *stochastic* in statistics). What we can predict is the time it takes for half of these iodine-131 atoms to disintegrate into xenon-131.

This period of time is called the *physical half-life,* which for iodine-131 is about 8 days, but from isotope to isotope of various elements, the half-life can last from billionths of a second (astatine) to billions of years (selenium-82). If we start with 1,000 atoms of iodine-131, after 8 days there will be 500 left, and after another 8 days there will be 250. It is natural to think that when something decays, what it produces will have a shorter half-life, but that is not always true. Sometimes what it produces has a much longer half-life.

Consider what happens within the core of a nuclear reactor like those at Chernobyl and Fukushima. The fuel rods are made up mostly of uranium-238 but also enriched uranium-235 that is about 4 percent of the total mass. When a neutron from a source—say, from the nucleus of another uranium-235 atom—strikes the nucleus of a new uranium-235 atom, additional neutrons are released. These can go on to strike other uranium-235 atoms in the nuclear fuel, in a chain reaction. When one of those neutrons crashes into the nucleus of an atom of plutonium-240, the captured neutron creates the radioactive isotope plutonium-241, which has a half-life of 14 years. But then the newly created plutonium-241 atom immediately begins to decay into americium-241, which has a half-life of 432 years. So of the radiation that comes from nuclear waste—spent fuel rods—some is from americium-241, not from plutonium-241; plutonium-241 will decay to an insignificant level in 140 years, but the americium-241 will be radioactive for over 4,300.

When radioactivity is introduced into a living organism—like us—another type of half-life comes into play: *biological half-life.* This is the amount of time it takes for the body to eliminate half of any material, radioactive or stable. When a radionuclide enters us, the amount of radiation we are exposed to (the *effective dose*) is a complex interaction between the physical half-life of the radionuclide and its rate of elimination from the body or other factors.

Biological half-life is complicated. It depends on several variables, such as the physical and chemical form of the radionuclide. For example, a gas like xenon may be inhaled, dissolve in the blood, and be excreted through the lungs. Xenon is one of the 6 "noble" gases that rarely react with chemicals in the body, so it passes through without harm. In contrast, strontium-90 interacts with chemicals in the body and can be incorporated into bone, because to the body, strontium-90 resembles calcium. Our bones are more dynamic than most people think: new bone is constantly being made, and old bone is constantly being reabsorbed. However, this process takes time. Consequently, strontium-90 will remain in our body far longer than xenon-131.

Adding to the complexity of biological half-life are the different potential routes of elimination, including breathing, urine, bile, and feces. A radionuclide that is eliminated mainly in the urine will stay in the body much longer in a person with kidney failure than in a person with normal kidney function. Because some radionuclides are also heavy metals, they may directly damage the kidney by chemical, nonradiation-related mechanisms. This means the body will be exposed to more radiation than it would were kidney function normal.

However, biological elimination can sometimes get radionuclides out of the body more quickly. For example, because cesium-137 resembles normal potassium, some of it is excreted in our sweat. Some scientists suggested putting victims of the Goiânia accident into a sauna to make them sweat more and get rid of the cesium-137 more quickly. Whether this would work or not is unknown. It is also possible to give victims of radiation poisoning chemicals that bind radionuclides efficiently, which are then excreted in the urine. British anti-Lewisite (BAL), used in World War II as an antidote to the chemical warfare agent Lewisite (a form of arsenic) and later as a means of removing heavy metals, such as lead and arsenic, from the body, is an example of such a chemical.

Combining biological elimination of a radionuclide with physical decay results in a faster loss than would occur in either process alone. Naturally, we are interested most in the interaction of ionizing radiations with human beings and with the plants and animals we eat: *How long does the radiation stay in me?* The answer depends less on the length of the physical half-life than on the effective half-life, which includes the biological half-life. Because of the disparity between physical half-life and biological half-life, our concern for the environmental impacts of a release of radionuclides may be entirely different from our concern for its direct health effects on people.

Let's translate the potential consequence of the physical half-life of a radionuclide on people's health. Assume someone inhales or ingests exactly the same number of atoms of two radionuclides with the same biological half-lives but different physical half-lives, radionuclide A and radionuclide B. Radionuclide A has a physical half-life of 1 second, and radionuclide B has a physical half-life of 400 years. For radionuclide A, most of the radioactive decay will occur within

10 seconds of entering the body. However, for radionuclide B, very few atoms will decay in these 10 seconds. Consequently, much less energy will be deposited in our body in a short interval. Because most of us live only 70 to 80 years, very few of the atoms of radionuclide B that enter our bodies can cause adverse health effects in our lifetime.

Alas, this is oversimplified. In reality, different radioisotopes emit different radiations. Some decays release gamma rays, which can travel a great distance from the atomic nucleus that released them, so energy from an atom disintegrating in your toe could possibly reach a cell in your brain. Other atomic decays release alpha particles, and because alpha particles are densely ionizing (at least when compared to an electron), they deposit their energy over a very short distance in the body. So an alpha particle released from a radionuclide in your toe may cause local damage but cannot affect cells in your brain. Thus when we consider the health effects of inhaling or ingesting a radionuclide, we have to consider the physical half-life, the distribution of the radionuclide in the body, and type of radiation emitted.

Let's consider cesium-137, which has a 30-year physical half-life but a biological half-life of 70 days in the human body. This means most of the atoms of cesium-137 that we inhale or eat are gone from our body in 700 days, or about 2 years from the date of intake. Why? To our bodies, cesium-137 chemically looks like potassium and is treated in much the same way, so it is easily absorbed and then constantly excreted in sweat, saliva, and urine. Strontium-90 has a physical half-life of 29 years, almost the same as cesium-137, but the body thinks that strontium is calcium, which is not easily absorbed. The 70 percent that is not absorbed and quickly passes out of the body in urine and feces; that fraction does not enter our health picture at all. But the remaining 30 percent gets into our bones and stays with us. Strontium-90 has a biological half-life of 49 years, so if you inhale or eat some strontium-90 at age 1, by the time you are 70 you will still have roughly a quarter of these atoms in your body. The strontium has gone through 2½ physical half-lives but less than 2 biological half-lives, whereas the cesium would have gone through 350 biological half-lives and been gone from your body for 68 years.

Radioactive decay continues until a radionuclide reaches a stable form. That can be a long and winding road, such as the natural decay chain of uranium-238, which in this order:

- decays, through alpha emission, with a half-life of 4.5 billion years to thorium-234
- which decays, through beta emission, with a half-life of 24 days to protactinium-234
- which decays, through beta emission, with a half-life of 1.2 minutes to uranium-234
- which decays, through alpha emission, with a half-life of 240,000 years to thorium-230
- which decays, through alpha emission, with a half-life of 77,000 years to radium-226
- which decays, through alpha emission, with a half-life of 1,600 years to radon-222
- which decays, through alpha emission, with a half-life of 3.8 days to polonium-218
- which decays, through alpha emission, with a half-life of 3.1 minutes to lead-214
- which decays, through beta emission, with a half-life of 27 minutes to bismuth-214

- which decays, through beta emission, with a half-life of 20 minutes to polonium-214
- which decays, through alpha emission, with a half-life of 160 microseconds to lead-210
- which decays, through beta emission, with a half-life of 22 years to bismuth-210
- which decays, through beta emission, with a half-life of 5 days to polonium-210
- which decays, through alpha emission, with a half-life of 140 days to lead-206, which is a stable nuclide.

To make this even more convoluted, some radionuclides decay through several different paths. For example, approximately one-third of bismuth-212 decays by emitting an alpha particle to thallium-208, while approximately two-thirds of bismuth-212 decays by emitting an electron to polonium-212. Both the thallium-208 and the polonium-212 are radioactive daughter products of bismuth-212, and both decay directly to lead-208, which (like lead-206), is stable and does not undergo further atomic decay.

What these drawn-out and arcane journeys of uranium-238 and bismuth-212 ultimately show is that as circuitous as half-life decay may be, and no matter how many millions or billions of years it may last, a radionuclide finally ends in usually one of two places. Over the very, very, *very* long run, almost all radionuclides become lead or iron.

A final complication is repeated exposure. The examples we've discussed up to this point assume one instantaneous exposure, like the atomic bomb explosions at Hiroshima and

Nagasaki, in which all of the energy and radiations from a bomb were released immediately. Contrary to what we might think, when an atomic bomb is exploded in the air, depending on the detonation height, it results in little radioactive fallout at the site, because the radionuclides it releases are sucked up into the mushroom cloud and injected into the lower atmosphere (troposphere). The radionuclides are then carried by winds to distant sites, including the poles. But an accident at a nuclear power facility like Chernobyl or Fukushima has a rather different and much more complex exposure scenario. Some radionuclides are deposited directly on the ground around the facility. Others are released into the atmosphere and cause fallout at different distances, depending on whether they are particles or gases and how large or small they are.

IODINE-131

The endocrine system encompasses the brain, ovaries, testes, stomach, pituitary and adrenal glands, and many more organs and tissues. The adrenals, for example, are small, triangular-shaped glands atop the kidneys that provide the boost of adrenaline released by surprise, excitement, or fear that is evolutionarily useful for fight or flight. All endocrine glands deliver the hormones they produce directly into the blood or to nearby cells rather than through a duct.

The thyroid is one of the largest glands of the endocrine system. It is a butterfly-shaped mass, below the Adam's apple, that produces the hormones that regulate metabolism, the chemical reactions that sustain life: it is a control panel of the human body. Iodine-127 enables the thyroid gland to perform these functions. Iodine-127 is thus vital to human health, but its cousin iodine-131 is a potential threat. Iodine-131, released as a gas or a particle, travels in the wind and can be inhaled. When deposited on grass, it is eaten by grazing cows and incorporated into their milk, a major source of food for children. Iodine-131, when it enters the body, concentrates in the thyroid gland. It can also get to the thyroid gland when a person drinks contaminated water or eats leafy vegetables contaminated on the ground. If your thyroid is not naturally topped up with iodine-127 when you are exposed, it will absorb the iodine-131 that you have inhaled, drunk, or eaten, because the thyroid cannot distinguish between the two iodines. The ionizing radiation of iodine-131 can cause a mutation in the DNA of a thyroid cell, and cancer can result. A simple way to prevent the thyroid gland from absorbing iodine-131 is to ensure that the thyroid is filled to capacity with regular iodine. Iodine-131, like a car without a parking space, then moves on with little or no effect, because all the places to capture iodine are filled with the nonradioactive form.

The best immediate intervention for persons who may be exposed to high doses of iodine-131 is to avoid consuming milk, milk products (like cheese), and fresh vegetables from the contaminated area and to take tablets of nonradioactive iodine, potassium iodide, *before* encountering iodine-131. However, these food precautions apply only to the first 80 days after a release of iodine-131 (10 half-lives). Thereafter essentially all the iodine-131 released will have decayed into stable xenon-131. It may not seem logical that a piece of cheese that was radioactive 1 month after a release of iodine-131 will not be radioactive 3 months later, but this is so. If the release of iodine-131 is not simply a one-time event but is ongoing, food precautions must be extended.

Only persons within the area of greatest exposure risk should take iodine tablets, especially children. This is because iodine tablets can also have adverse health effects. Many people are allergic to iodine; those with some thyroid diseases may worsen after taking too much iodine. And children especially may be accidentally poisoned by taking too much iodine. The risks from taking too much iodine are not on the same scale as having too little. Still, taking too much can lead to toxic effects, including, in some people, overactive thyroid gland (thyrotoxicosis), a disease that can be fatal.

In the days following the Fukushima explosions, it was almost impossible to find potassium iodide tablets in drugstores in California, 5,000 miles away, because people had bought them up. This was a pointless precaution. Californians taking iodine tablets to protect against iodine-131 released in Fukushima would have been just as well served if they had bought raincoats to protect against showers in Barcelona. We do not mean to discount the potential harmful effects of a dose of iodine-131 to someone in its immediate path. But the effect of released iodine-131 can be tempered by both fortune and reaction time. The danger to people from the iodine-131 released in the Chernobyl reactor meltdown in 1986 was strikingly dissimilar from that released in the Fukushima explosions in 2011.

At the time of the Fukushima disaster, the winds were blowing eastward—offshore—and carried most of the iodine-131 out to sea. (Bob was soon at the scene of the accident.) But even if the wind had instead blown over land, the danger would not have been as grave as one might think. The Japanese do not have an iodine-deficient diet—they regularly receive a large amount of iodine from fish, seaweed (kelp is naturally high in potassium iodide), and other sources. In contrast, the farmers in Belarus, Ukraine, and Russia had to subsist on locally produced milk, milk products, and vegetables, before and after the Chernobyl accident, but Japanese children ate uncontaminated food that was brought in from elsewhere. Milk is a major source of iodine, and kids get more of their nutrition from milk than do adults. This is important because young people, whose thyroids are larger compared to their body size than adults' thyroids, tend to concentrate more iodine in their thyroid gland and so receive a much higher radiation dose there; infants are 5 to 30 times more susceptible to a dose of iodine-131 than adults. This age-dependent risk of thyroid cancer is quite important in estimating a person's cancer risk. Most of the excess cancers in the A-bomb survivors and in the populations exposed to radiations from Chernobyl were in children and adolescents who were younger than 20 years old when the exposure occurred. Adults are relatively resistant to the thyroid cancer–causing effects of radiation.

In most of the severely affected areas in Japan, potassium iodide tablets were effectively distributed and people were evacuated. Measurements of radioactive iodine in 1,000 children from Fukushima prefecture show very low levels or none. The Japan Academy of Science (the equivalent of the U.S. National Academy of Sciences) predicts that few, if any, thyroid cancers will result from iodine-131 at the Fukushima Daiichi accident.

Preliminary external radiation measurements have now been made on about 175,000 Japanese after the Fukushima accident. They suggest relatively low levels of internal exposure. Based on these data, the major health consequences of the Fukushima accident are likely to be psychological, from the enormous loss of life and social and economic disruption caused by the earthquake and tsunami. A decades-long study of the effects of radiation on thyroid disease in children living in the most exposed areas around Fukushima has begun; they will have ultrasound studies of their thyroid every two years from age 2 to 13. One problem with a broad study like this is that there are no data on normal children of this age not exposed to radiation. The study may detect many abnormalities that have nothing to do with radiation exposure. Some of these abnormalities will undoubtedly lead to biopsies and other interventions, some or many of which may be unnecessary or even harmful. (This complexity of attempting to balance potential risks and benefits is similar to current debate over screening for breast and prostate cancers.) Moreover, in contrast to studies in the A-bomb survivors, we have no control group of unexposed children, such that any conclusion must rest on a comparison of children receiving lower versus higher doses. Finally, we lack convincing data that early detection of thyroid cancer results in a health benefit.

After the Fukushima accident, some proposed that the Fukushima emergency workers should have blood or bone marrow cells collected and frozen, in case they developed radiation poisoning of the bone marrow and needed those cells for a transplant. These blood cells could be collected before the high-dose exposure occurred. But others opposed the idea, arguing that there are more than 20,000 Fukushima workers, and no one had an idea which ones might accidentally receive a high dose of radiation; collecting blood cells from so many people is associated with substantial health risk; and it was highly likely these cells would never be used. This difference of opinion led to considerable controversy among Japanese specialists. Eventually Bob and his Japanese colleagues persuaded the government and the Japan Academy of Science that this was a bad idea. (More than a year after the accident, it was alleged that some workers involved in the aftermath of the accident were ordered to cover the dosimeters, which measured the amount of radiation they were exposed to, with lead to prevent their recording the actual amount of radiation received. This, if true, was unconscionable.)

After the Chernobyl accident, Bob and his Russian colleagues decided to transplant 13 emergency personnel who received extremely high doses of radiation and were unlikely to survive the destruction of the bone marrow. Even there, the complexity of injuries and damage to tissues and organs other than the bone marrow led Bob and his colleagues to conclude that transplants were likely to be useful in only a very small proportion of nuclear accident victims.

In the weeks following the Fukushima accident, there were reports that milk from the area exposed to radiation released from the nuclear plant had elevated amounts of iodine-131. Because of its possible immediate harmful effects, a large quantity of milk was destroyed.

Other spikes in radiation detected in the weeks and months following the Fukushima accident are best understood more broadly to gauge their real danger. On the day that a higher-than-permissible level of iodine-131 (300 becquerels per liter, or less than 0.00008 parts per *trillion*) registered in the drinking water in Tokyo, residents would have had to drink 6 liters a day for a month to receive an internal dose of radiation similar to the external dose an airline crew receives in a year flying between Los Angeles and Tokyo. Few data show convincing increased risk of cancer among airline crews, and any increase there may be is low. This issue is, however, controversial, and more breast cancers are reported in female airline crews than in other groups of women who fly less. Iodine-131 can be measured so precisely that regulators can set a limit of parts per trillion; cautionary levels for other substances more difficult to measure are in parts per million or billion; the measurement of iodine-131 is thus 1,000 to 100,000 times more sensitive.

Were fish from the ocean around Fukushima safe to eat following the discharge of radioactivity into the sea? This was a natural worry, and stringent mitigation actions were taken. The real issue is, how can we be certain any food source is safe if higher than normal radiation is detected? The best way to frame the discussion is to consider the risk involved with eating food that may be contaminated by radionuclides.

F. Owen Hoffman, president of the SENES Oak Ridge Center for Risk Analysis in Oak Ridge, Tennessee, is a leader in the field of radiation risk assessment. He is a distinguished emeritus member of the National Council on Radiation Protection and Measurements (NCRP) and consultant to the United Nations Scientific Committee on the Effects of Atomic Radiation (UNSCEAR). In June 2012 Hoffman wrote to us that "situations like this cause health agencies to want to declare such contaminated samples that enter the marketplace as either 'safe' or 'unsafe,' without full disclosure of the measured amounts of radiation involved."

> Another dilemma is that often [U.S.] food safety standards restrict carcinogens to levels at about one in one hundred thousand or one in one million lifetime risk of cancer incidence in later life. For a person of average age and gender (there is no such thing), these regulatory risk levels for carcinogens would amount to cumulative whole body lifetime doses of a mere 0.01 to 0.10 mSv, dose levels far below what any radiological health organization would deem as clearly "safe."

Our bodies can tolerate a certain amount of ionizing radiation, perhaps more than most people think. Consider the releases of radioactive iodine-131 and cesium-137 from Chernobyl, from Fukushima, and from the atmospheric atomic weapons tests carried out in the 1950s and 1960s. Chernobyl is estimated to have released 5 to 10 times more radioactive materials than Fukushima, although not everyone agrees with this estimate, and data are presently incomplete. The atmospheric weapons tests released about 200 times more radioactive materials than that released at Chernobyl, and 2,000 times more than that released at Fukushima. (These estimates are controversial and are meant only to give a possible scale for comparison.) The greater magnitude of the Chernobyl releases was due to many reasons but especially the lack of an effective containment structure typical of most commercial nuclear power reactors.

Estimated Radionuclide Releases of Radiation

	iodine-131 (EBq)*	cesium-137 (PBq)†
Chernobyl	1.8	80
Fukushima	0.15	13
A-bomb	675	950

*EBq=exabecquerel (a quintillion—10 followed by 18 zeros, or a billion, billion
†PBq=petabecquerel (a quadrillion—a million billion)

Exposure to radiation isn't always what it seems. Victims of the A-bombs in Hiroshima and Nagasaki received almost all their exposure in an instant, a point we made earlier. But when a nuclear power facility accident deposits radionuclides into the environment, nearby residents can be exposed throughout their remaining lifetimes, assuming they are not relocated.

Of course, people in the path of a radioactive cloud may receive a dangerous dose of radiation, depending on the concentration of radionuclides, atmospheric conditions, and whether they are indoors or outside when the radioactive plume passes. Immediate countermeasures taken to protect the public are quite important. The Japanese government's response to the Fukushima accident, well-conceived or not, was quite effective. Most people were sheltered and then evacuated in a relatively controlled fashion. Some received iodine tablets. However, luck may have played an important part as well. Very fortunately, the predominant offshore winds blew about 80 percent of the released radiation into the Sea of Japan. (We will explain in chapter 8 why this was fortunate and not horrendous.) When an independent commission reviewed documents related to the Fukushima accident, however, it found considerable confusion between the Japanese government, officials at the nuclear power facility, and executives at the Tokyo Electric Power Company (TEPCO) headquarters. Furthermore, emergency authorities did not share or use some important data on radioactive contamination, such that some people (fortunately few) were evacuated to zones of higher, rather than lower, contamination. And some children remained in high radiation areas for far too long. Still, the fortunate bottom line of these effectively planned if sometimes confused actions is that they were largely successful in protecting the public.

This contrasts with the Chernobyl accident, where winds were blowing toward Europe and Scandinavia, with no intervening large body of water in which to dilute the released radioactive materials. Also, the infrastructure of the Soviet Union was rather different than that of Japan. Soviet authorities rapidly evacuated about 40,000 people living in Pripyat, the high-rise city less than two miles from the crippled reactor. Residents also received iodine tablets. However, evacuation of people living in a 20-mile exclusion zone was more complex because of less efficient communications. Soldiers went door to door for several days to persuade residents to leave. Because many people lived on self-sustaining small farms, it was impossible to quarantine milk and milk products and locally grown vegetables, which are the main components of the residents' diet.

The radioactive cloud moved over rural areas filled with these small farms. The people living in the cloud's path had no alternative source of food. As we said, they had to eat what they grew and drink milk from their cows. In addition, the lack of infrastructure made it difficult or impossible to provide iodine tablets to those in the path of the cloud, except those living in Pripyat. Many people who live near U.S. nuclear power facilities now have iodine tablets in their homes in case there is a release of radioactive gases.

From a media viewpoint, the Soviet government's response to the Chernobyl meltdown was glacial compared to the Japanese response to Fukushima. The explosion in reactor 4 of the Chernobyl nuclear power facility took place on April 26, 1986, but it was May 14 before the first official announcement, by which time the radioactive plume had passed over Belarus, Russia, Ukraine, and most of Western Europe and Scandinavia. It is estimated that 60 percent of the radiation settled on Belarus and the millions of people living on the produce of their family farms.

Many believe hundreds of thousands of people have died from illnesses caused by the radioactive releases from the Chernobyl accident (a recent news item in an Australian newspaper placed the number at 1 million) and that there has been a huge cover-up. It is easy for some to place responsibility for illnesses and death on a catastrophic event rather than on the vicissitudes of life. The World Health Organization reported in 2006 that in the areas most heavily contaminated by radioactive fallout from Chernobyl, the number of deaths from cancers was no higher than normal for this population. However, because no iodine tablets were distributed, and because so many children unwittingly drank milk contaminated with iodine-131, in Belarus, Russia, and Ukraine over the three months following the accident, there are more than 6,000 cases of thyroid cancer, and of those victims, according to UNSCEAR, about 15 have died. Other short-lived radioactive forms of iodine such as iodine-128 may also have played a small role in the development of thyroid cancers, as may cesium-137. But iodine-131 seems to be the major if not sole culprit. Recent studies show that thyroid cancer risk remains elevated more than 25 years after the accident. This level should begin to decrease over the next decades. Total estimated thyroid cancers may be more than 10,000 and as many as 16,000. Fortunately, most thyroid cancers are cured.

How many more cancers will occur as a consequence of radiation released by the Chernobyl accident? Estimates range from the 6,000 cases of thyroid cancer in children and adolescents to several hundred thousand other types of cancers in adults. A recent estimate from the International Agency for Research on Cancer predicts as many as 25,000 cancers by 2065. Although this is disturbing, we need to recall that several hundred million people will develop cancer from other causes during this interval. The correct number of Chernobyl-related cancers will never be known, in part because of the considerable uncertainties in estimating cancers and cancer deaths. Especially problematic is the controversy about whether very low doses of radiation, especially if given over a prolonged interval, increase cancer risk.

There are other difficulties as well. For one, we do not know precisely what radiation dose most people received. People who were indoors when the radioactive plume passed received much less radiation than those who were outdoors.

However, because most people did not know when the radioactive plume passed, they cannot accurately reconstruct their whereabouts at that time. Also, many people were evacuated from contaminated land at different times and thus received very different doses from ground and food contamination.

Next we have the geopolitical reality that many of the exposed people no longer live in the Chernobyl area. Living elsewhere, even in other countries, they are lost to follow-up. The Chernobyl accident was relatively quickly followed by the dissolution of the Soviet Union, whereupon many people's lifestyles but perhaps not their lives changed, mostly for the worse. For example, cigarette smoking and alcohol consumption increased, resulting in a profound drop in life expectancy. Both activities are correlated with increased cancer risk independent of radiation exposure. Sorting out any changes in cancer incidence or prevalence will be difficult at best. Many scientists think even a reasonable estimate of health consequences of the Chernobyl accident for populations other than the 600,000 or so mitigation workers (called "liquidators") is impossible.

Recall that the estimated lifetime cancer risk for females is about 38 percent and for males about 45 percent. About one in every three women and almost half of males will develop cancer in their lifetime. Even a large number of Chernobyl-related cancers, say 100,000, would alter the baseline cancer risk by less than 0.1 percent. Given these considerations, it is easy to understand why there is so much controversy about the late consequences of the Chernobyl accident. We should also point out that most cancers attributed to radiation from the A-bomb explosions (about eight percent of all deaths) were detected several decades later and that it has been only about 25 years since the Chernobyl accident. This is a very brief period in the context of radiation-induced cancers, save for leukemias and thyroid cancer.

Because people did not understand the dangers of ionizing radiation, an estimated 100,000 abortions were performed after Chernobyl on women in the former Soviet Union and Europe. Mothers and doctors incorrectly thought these fetuses were at risk for birth defects from radioactive fallout. These abortions were unnecessary. In Japan, about 3,000 pregnant women were exposed to high doses of radiation from the A-bombs, but only 30 children had detectable birth defects. At the time of the blasts, all the children with birth defects were fetuses in their second trimester of development—a critical time, when nerve cells migrate from their embryonic location; radiation exposure seems somehow to have interfered with this normal migration. So more than two-thirds of the fetuses exposed to doses of radiation from the A-bombs—substantially higher doses than anyone in Belarus, Russia, Ukraine, and Europe could have possibly received from Chernobyl—were unaffected. And no pregnant woman's fetus was hurt by the fallout from Chernobyl. (There were no pregnant women among the firefighters, rescue workers, or the thousands of liquidators involved in the cleanup.)

Birth defects, including the subtlest and thus most difficult to detect, are found in up to 10 percent of children in the "normal" U.S. population. When someone who lived in Ukraine had a child with a birth defect, it did not necessarily result from radiation released at Chernobyl. Moreover (as we will discuss in greater detail in chapter 5), radiation-induced genetic abnormalities are not passed from the affected persons to their children, as studies of exposed Japanese mothers and their children make clear.

At Chernobyl, there was a nearly complete meltdown and most of the fission products in the reactor core were released. The nuclear fuel was completely exposed as the chain reaction was ongoing, so neutralizing boron (a "neutron poison," so called because it absorbs many neutrons and stops a chain reaction) was dumped over it from helicopters. After eleven days of work, the risk of further emissions was largely eliminated, though by then, of course, tremendous damage had been done. In Fukushima, at this writing, there is not yet full access to compromised reactors. Assuming there is no further catastrophic damage—which is not a given, though more promising as time goes by—the danger of one or more critical events is decreasing. However, it will take several decades to decommission the reactor complex, for safety reasons and because not all the necessary technology for this task currently exists. It also makes sense to proceed slowly to allow for decay of some of the shorter-lived radionuclides.

EFFECTS OF FALLOUT FROM ATMOSPHERIC TESTING OF NUCLEAR WEAPONS

From 1951 through mid-1962, atmospheric nuclear weapons tests at the Nevada Test Site released a great deal of iodine-131 into the environment in the United States and elsewhere. More was released as a result of the underground tests that came after the Limited Test Ban Treaty of 1963. The U.S. National Cancer Institute (NCI) analyzed the doses of iodine-131 that the about 170 million Americans living in the continental United States received during the years of testing. The NCI's findings on thyroid cancer are important. People's risk of developing thyroid cancer from iodine-131 released by these tests depends on how old they were when they were exposed, where they lived, and perhaps most important, how much and what type of milk they drank. Someone who drank one to three 8-ounce glasses of milk from a backyard cow or a goat got a dose of iodine-131 that was six to sixteen times higher than if they drank an equal amount of commercial cow's milk. (Goat's milk has the highest amount of iodine-131—yet another example of varying outcomes where radiation is involved.) On the other hand, a child who was breast-fed during its first year of life received a radiation dose to the thyroid about 30 percent lower than a child who drank commercial cow's milk. Children younger than 20 at the time of the tests were at highest risk of radiation-induced thyroid cancer, because a child's thyroid concentrates more iodine-131 than an adult's, so the radiation dose to the thyroid is higher. Women living in the same region as men had a three times greater chance of developing thyroid cancer. Around 50,000 new cases of thyroid cancer are diagnosed in the United States every year, usually in women between the ages of 25 and 65. With treatment, most are cured; there are only about 1,500 deaths annually.

Before the long-term effect of radiation on the thyroid was understood, children who emigrated to Israel in the 1940s and 1950s received external X-ray radiation treatments for ringworm. This led to an increased risk of thyroid cancer and brain cancers decades later. And some American children who received similar external radiation for an enlarged thymus gland (erroneously thought to cause increased infections) or tonsils in the 1940s and 1950s also developed thyroid cancers. Thyroid cancer can occur years later in people who receive external radiations to treat Hodgkin lymphoma or to prepare for a bone marrow transplant.

For all the dangers that concentration of iodine-131 in the thyroid can cause, an intentional high dose of iodine-131 can sometimes have benefits. Doctors use radioactive iodine-123 or iodine-131 to diagnose thyroid abnormalities such as under- or overfunction (hypo- or hyperthyroidism) or to determine if a thyroid nodule is metabolically active (concentrates iodine). Nodules that take up iodine-131 (hot nodules) are less likely to be a cancer than cold nodules. Diffuse uptake of a large amount of iodine-131 is characteristic of hyperthyroidism. If a thyroid nodule is determined to be a cancer (usually after a needle biopsy), the first step is surgery, unless the cancer has spread. Doctors sometimes give very high doses of iodine-131 to selectively kill thyroid cancer cells.

Iodine-131 illustrates an interesting paradox of radiation-induced cancers: more is not always worse. Low doses of iodine-131 are shown to cause thyroid cancers in several settings we discussed, but very high doses of iodine-131 rarely cause thyroid cancer. Why? And how does this fit with the linear no-threshold radiation-dose concept? The reason for the paradox is that very high doses of iodine-131 kill normal thyroid cells and a dead cell cannot cause cancer. But a low dose of radiation can cause sufficient mutations in DNA for a normal cancer cell to survive and become a cancer. Consequently, doctors usually give only very low doses of iodine-131 (as in thyroid uptake and scan studies) or very high doses of iodine-131 (as in treating thyroid cancer) but rarely intermediate doses.

People exposed to ionizing radiation are sometimes the target of discrimination based on fear that they will contaminate others. The Japanese term for survivors of the A-bomb blasts of 1945 is *hibakusha* ("explosion affected people"). Some *hibakusha* are reluctant to admit who they are for fear of discrimination against themselves or family members. It may be harder for the child of a *hibakusha* to find a marriage partner because of the parent's radiation exposure. In the case of the Goiânia incident, relatives who drove to Rio weeks afterward to see their family members or friends in the naval hospital experienced blatant discrimination. Their car license plates showed where they came from, and because some in Rio feared that these visitors brought radioactive materials from Goiânia with them, signs were placed on their windshields telling them to go home, and in several instances their cars were vandalized. However, the workers who cleaned up Chernobyl were given special benefits by the government and are admired by their fellow citizens. This fear of persons exposed to radiation is reminiscent of people's fear of lepers during the Middle Ages and even today. Leprosy is caused by a bacterium and is notoriously difficult to "catch" from another person unless you are a lifelong companion. Some cured lepers, however, chose to remain in "colonies" because they feared discrimination. Lest we think fearing lepers is a strictly medieval mentality, recall the early days of the AIDS epidemic. The lesson is that ignorance of science can lead to irrational behaviors that harm others and ultimately ourselves.

Some argue that governments should sponsor screening of potentially exposed populations for cancer or radiation, but experience has shown that such screenings are not always necessary or desirable and may have unanticipated consequences. After the collapse of the Soviet Union in 1991, more than one million immigrants came to Israel, some from areas affected by the radioactive cloud from Chernobyl. The Israel Ministry of Health debated whether to set up a special clinic to screen those émigrés for cancer. Bob was consulted at one point. Could possible early cancer detection be a benefit?

There are several important considerations in determining whether such a screening program is worthwhile. First, will cancers be increased to any great magnitude in the exposed population? Second, will early detection be of benefit? (Often the answer is not as obvious as it may seem.

Consider, for example, the controversies we mentioned over early detection of prostate and breast cancers; and see chapter 6.) Third, what potential adverse effects could result in the screened population? Might the screened people reasonably develop a fear of developing a radiation-induced cancer, without actually developing one? Sometimes the magnitude of this fear may outweigh any potential early detection benefit. Israel reasonably chose not to institute a screening program for these immigrants from the former Soviet Union.

THE AFTEREFFECTS OF CHERNOBYL

It is now more than 25 years since the Chernobyl nuclear power facility accident. We have so far discussed the diverse health effects of acute exposure to high-dose radiation on the facility workers and the emergency personnel who responded to the accident, as well as thyroid cancer among children exposed to iodine-131.

The next group to consider are the roughly 560,000 "liquidators" who helped control the releases, who cleaned up and/or decontaminated the surrounding areas, and who built the concrete sarcophagus in which the remains of reactor 4 are now entombed. Several national and international organizations, including the United Nations and the International Agency for Research on Cancer, have sponsored studies of cancer risk in these people. These studies have tried to follow the health of these workers for twenty-five years, but as we mentioned, the studies have been difficult to conduct for several reasons.

In brief again, first, the Soviet Union dissolved, so researchers had to deal with different countries and with health ministries in those countries. Second, precise data on exposures for many of the workers was lacking. Third, many people were displaced and lost to follow-up. Fourth, high-quality cancer registries were absent before and even after the accident, making it impossible to know with certainty the background rate of most cancers before the accident.

Despite these limitations and others, it is important to try to quantify radiation-related health consequences of the Chernobyl accident for several reasons. As we've said, A-bomb survivors and people receiving medical interventions are our main source of data regarding effects of ionizing radiations on humans. However, these populations were exposed to high doses of radiation instantaneously. Data from Chernobyl have the potential to show the health consequences of lower doses of radiation received over a long interval.

The most striking delayed radiation effect of the Chernobyl accident, as we saw, was the more than 6,000 cases of thyroid cancer in children and adolescents, the consequence of ingesting large amounts of iodine-131, mostly via the food chain.

But what about other cancers? Here the situation is less certain. There are some reports of an increased incidence of leukemias and/or hematological cancers like lymphomas (a lymph node cancer) and multiple myeloma (a bone marrow cancer) in the liquidators. Fortunately, the size of these increases is small, and the validity of the findings is debated. However, on average, the relationship between radiation dose and size of the increased incidences of these blood cancers is compatible with data from the A-bomb survivors. But because the radiation doses that the cleanup workers received at Chernobyl were so much smaller than doses received at Hiroshima and Nagasaki, the number of excess cancers is much smaller. This is good news for the larger exposed population, including about 200,000 evacuees and relocated persons and people still living in contaminated parts of Ukraine, Russia, and Belarus who, on average, received far lower radiation doses than the liquidators. The small, if any, increase in leukemias is also good news in that it suggests there will be relatively few solid cancers over the next decades. Except for thyroid cancer in children and adolescents, the very small increase in cancers from Chernobyl is unlikely to be detectable and would increase this background cancer risk by less than 1 percent.

To understand (and believe) this, we need to compare the radiation doses received by these populations to our normal radiation dose. Americans are exposed to about 6.2 mSv each year, so in 20 years we will receive about 125 mSv. Among the Chernobyl populations, the 560,000 or so liquidators received an average dose of 100 to 200 mSv, or about what most Americans receive over 20 years. The dose to the evacuees was 30 to 50 mSv, or about one-quarter to one-half of the 20-year American dose. The quarter-million residents of lands with high ground contamination received about 50 mSv, and the 5 million people living in lands with low ground contamination received 10 to 20 mSv. So when we put these Chernobyl-related doses in context, it is reasonably clear that a large increase in cancers is unlikely.

Now, let's compare late adverse health effects caused by the Chernobyl accident and by the Hiroshima and Nagasaki A-bombs. As we've mentioned, the joint Japan-U.S. commission to study health effects of the A-bombs started collecting data in 1950, so there are no precise data on earlier events. However, beginning in 1950 and perhaps earlier, deaths from leukemias increased dramatically. This increase continued until 10 to 15 years after the A-bombs, when numbers of excess cases of leukemia declined, eventually to baseline levels.

In striking contrast, an excess of several other cancers—including stomach, lung, liver, colon, breast, gallbladder, esophagus, bladder, and ovarian—increased slowly over the next several decades and remained elevated throughout life. The size of the increase is proportional to the estimated radiation dose received. Interestingly, not every type of cancer increased. For example, there was no detectable increase in cancers of the rectum, pancreas, uterus, prostate, or kidney. Are some cancers less radiogenic than others, or was the increase of some cancer types too small to be detected in a sample of this size? We do not know. Both explanations are possible. The risk of developing radiation-related cancer was greatest in persons who were youngest when they were exposed to radiation. The risk of death from heart and lung diseases and gastrointestinal disorders also increased in the A-bomb survivors, but whether this increase resulted from radiation exposure is uncertain.

Despite these strong scientific data, stories in newspapers, magazines, and even books continue to describe children with birth defects supposedly from Chernobyl and, in one instance, a three-headed cow. These claims have no basis in fact.

A NEW TREATMENT FOR RADIATION POISONING

The treatment of some of the Chernobyl victims of radiation poisoning was a medical advance. Soon after Bob arrived in Moscow to help treat the firefighters and other victims of the accident, his UCLA colleague David Golde, who was working on the development of molecularly cloned granulocyte-macrophage colony-stimulating factor (GM-CSF, which was used in Goiânia the next year) suggested that this still-experimental drug might help the victims' bone marrow recover more quickly. If ever there was a time to speed research, this was it. Several of the injured were going to die of infection and bleeding unless their bone marrow could recover quickly. Bob's Soviet counterpart, Dr. Andrei Vorobiev, agreed that GM-CSF was the best hope. Their first task was to obtain a supply of the drug. If they succeeded, then they would deal with getting permission from Soviet medical authorities to use it.

Dr. Golde and his coworkers were developing GM-CSF with scientists at the pharmaceutical company Sandoz (which is now part of Novartis), in Basel, Switzerland. Bob called Dr. Angelika Stern, an acquaintance at the company, and asked for enough of the hormone to treat several victims with radiation sickness who had no transplant donor. Stern and Sandoz agreed. That left the problem of how to get the drug into the Soviet Union in the middle of a crisis about which the government was doing everything possible to limit information; this was in the midst of the Cold War, and secrecy was a given. Sandoz recruited a Swiss businessman en route to Moscow to carry a parcel packed in dry ice, without his knowing its contents. Bob and his Soviet medical colleagues made arrangements to alert the security guards at the airport that the package was intended for the doctors treating the Chernobyl victims, and there could be no delay. All went as planned.

The businessman was ushered quickly through border control and customs. As instructed, he called Bob, who picked up the package and went directly to the Clinical Hospital No. 6, a Kremlin-related hospital that was purposely out of the mainstream. The hospital was associated with the Institute of Biophysics (now the Burnasyan Federal Medical Biophysical Center), where all casualties of the Soviet nuclear program had been taken over the years. Its directors were Angelina Guskova, a specialist in the treatment of radiation sickness, and Lenoid Illyin. Guskova as a young physician worked closely with Dr. Igor Kurchatov (the Soviet counterpart of J. Robert Oppenheimer), who directed the development of the country's atomic bomb, first exploded in 1949. (Kurchatov contributed to building the first Russian hydrogen bomb in 1950, but in the years before his death in 1960, he advocated peaceful uses of nuclear technology.)

Getting permission from the Soviet authorities to use the drug proved more difficult than bringing it into the country. General Secretary Mikhail Gorbachev and the Politburo members were distracted by a worldwide public relations nightmare caused by their slow release of information, and they did not want another controversial topic on their agenda that could backfire. They refused permission to use the drug, claiming the excuse that they did not want to be accused of making the accident victims guinea pigs for an untried therapy, even though clinical trials were about to begin worldwide.

Bob and his Soviet colleagues were reasonably convinced that giving GM-CSF would not be dangerous, and substantial data in radiated animals, including monkeys, showed it rapidly improved bone marrow function. It might save some of the radiation victims. But how were they to overcome this mostly bureaucratic public-relations-oriented objection?

If the Soviets didn't want Chernobyl victims to be the first people to receive GM-CSF, Bob was willing to try it. He asked his Russian colleague Andrei Vorobiev to inject him with the drug, and added, "If I don't drop dead, we are over the 'first human' hurdle." Vorobiev, a member of the Academy of Medical Science whose official title was *Academician* (speaking to his influence) and a generation older than Bob, immediately agreed but also wanted to be injected with GM-CSF. That evening, after calculating the appropriate dose based on data from monkeys, they injected each other and agreed to meet at the hospital at eight the next morning, when a blood test would show whether their granulocytes had increased. If yes, this would be a sign that the drug might help the radiation victims.

After leaving the hospital, Bob went to Spaso House, the American Ambassador's residence in Moscow, to have dinner with U.S. Ambassador Arthur Hartman. Also in residence were a number of Soviet dissidents who had been granted asylum. Because of the political sensitivities, Bob usually kept his distance from U.S. government officers, and they did likewise with him. But the ambassador was eager to know what was going on, and Bob told Hartman everything he thought not confidential.

Midway through dinner Bob was called from the table to take an urgent phone call. "Academician Vorobiev," he was told, "has just been admitted to the hospital and is critically ill."

Bob's immediate, deadpan thought was, *If he dies, first, it will be rather unfortunate, and second, it is going to ruin our chance to help the victims.* He rushed to Hospital 6, and entering the room, he saw Vorobiev on his back in bed, his face colorless, complaining of chest pain. Everyone assumed Vorobiev had suffered a heart attack, but an EKG and blood studies showed nothing wrong. Bob questioned Vorobiev further and examined him. He concluded the severe pain was in his sternum (breastbone), not the heart. What the researchers could not have known from rats and monkeys is that when this hormone is given, blood vessels in the bone marrow contract and squeeze out granulocytes into the blood. There are many nerve endings in the bone marrow, and as Vorobiev learned, this process can be quite painful. (Bob did not have any pain, but now after years of use in humans, it is well known that the drug frequently causes severe pain in the sternum.) The next morning Vorobiev felt normal, and he and Bob gave blood samples. The granulocytes were much higher.

GM-CSF proved useful in treating the radiation victims, and it and related drugs are now given to tens of thousands of people with cancer who are receiving chemotherapy that suppresses the bone marrow, an effect much like that from radiation. It is also given to children with rare genetic diseases of the bone marrow and to normal people donating blood or bone marrow cells for relatives with cancer. This approach is now a standard intervention for radiation accident victims.

Of the people Bob and his Soviet colleagues treated, Andrei Tarmosian, a 26-year-old who survived, had an especially poignant fate. He was one of the firefighters who rushed into the burning reactor to try to put out the inferno. Although severely burned by the radiation, he eventually recovered and returned to his family. In time he became a grandfather. Bob keeps on his computer a picture of Andrei and his grandchild as an infant. They've corresponded by mail and then e-mail several times a year.

But because of the firefighter's close call, the specter of death hung over him. Despite his recovery, he was (wrongly) convinced he would eventually die of radiation-induced cancer. He drank heavily, both to alleviate his fears and because he believed the rather common Russian notion that vodka protects against radiation. Andrei had a very small chance of developing a radiation-related cancer, but his background risk was already about a 45 percent chance, as it is for all males. In 2010 Bob received a late Christmas card from Andrei's daughter, telling him that her father had died from cirrhosis of the liver (probably from drinking alcohol), age 50.

THE DIFFERENCE BETWEEN THE CHERNOBYL AND FUKUSHIMA ACCIDENTS

Commercial nuclear reactors that use normal, or "light," water have two basic designs. In one, heat generated from fissioning uranium boils water within the reactor core. This type is referred to as a *boiling water reactor* (BWR). In the second type, water entering the reactor core comes under high pressure, like water in a pressure cooker which prevents the water from boiling. It is referred to as a *pressurized water reactor* (PWR). The superheated water transfers its thermal energy to water outside the reactor core, which then boils. In both the BWR and the PWR, the boiling water produces steam, which is then used to spin turbogenerators to produce electricity. In this regard they are no different from coal-powered power facilities, which use the thermal energy from burning coal to boil water and produce steam, or from conventional solar power facilities, which use another radiation source, the Sun, to boil water to produce steam. The generation of hydroelectric power is in some ways similar, except that the turbogenerator blades are spun by the force of falling water rather than by steam. There is nothing particularly magical about how electricity is produced using nuclear energy, except that the amount of energy per unit mass of the uranium fuel—the energy density—is many thousands of times greater for uranium and plutonium than it is for hydrocarbons such as oil and coal. Most Western commercial nuclear reactors are PWRs, as are reactors in nuclear submarines.

An important aspect of commercial nuclear reactors is the *moderator.* Neutrons are released from uranium at too high a speed to sustain a chain reaction, and a moderator is needed to slow them down. All light-water reactors use the coolant water to slow the neutrons to a speed that will sustain a chain reaction.

The reactors at Chernobyl were an early iteration of a nuclear reactor called RBMK (*reaktor bolshoy moshchnosti kanalniy,* or high-power channel-type reactor), which has what some nuclear engineers consider a design flaw. RBMK reactors can be used to produce weapons-grade plutonium, which may be why they were developed during the Cold War. They use graphite control rods (graphite is a neutron-absorbing material) rather than the coolant water to slow down neutrons and sustain the chain reaction in the reactor core, although like other BWRs, they use water as the coolant. When a RBMK reactor loses coolant, the reactor core heats up. However, in contrast to a Western-type reactor, where the water is also the moderator such that the chain reaction slows down, in a RBMK reactor the chain reaction accelerates because the moderator, graphite, is still there to slow down the neutrons. A loss-of-coolant accident in an RBMK reactor is like an out-of-control car whose driver hits the gas pedal rather than the brake; the reactor too accelerates out of control. This is the opposite of what happens in a Western-type reactor, where water is both the coolant and the neutron moderator.

The Fukushima Daiichi reactors, newer (but not new) and better designed, were made by General Electric and are called Mark 1 reactors. A Mark 1 reactor uses water both as the neutron moderator and coolant. As a result, when there is a loss-of-coolant accident, as at Fukushima, loss of the moderator slows down the chain reaction and shuts off the reactor. This is like hitting the brake in an out-of-control car; the car slows and then stops. At Fukushima the reactor slowed down rather than sped up, but not enough.

Most experts agree that it was predominantly the tsunami, not the earthquake, that severely damaged the Fukushima facility, although some earthquake damage may also have occurred. Mark 1 reactors need circulating water to cool the reactor core. This requires electricity, which can come directly from the turbogenerators at the facility or from the regional electrical grid to which the facility is linked by high-voltage cables. If these electrical sources fail, diesel generators at the facility will kick in and generate electricity.

The earthquake knocked out electrical generation at Fukushima Daiichi and at all other facilities on the electrical grid. Next, the diesel generators started and began supplying electricity to the water pumps. But then came the final blow. An extraordinarily high wall of water (estimated to be about 45 feet) from the tsunami inundated the diesel generators, shutting them down. Now the reactors were at risk. The chain reaction in the reactor core continued to generate heat, albeit less, but there was no circulating water to remove this heat. Although concomitant loss of the neutron-moderating coolant water worked to slow the chain reaction down, it was not fast enough, and the nuclear fuel began to melt. Whether the reactor designer and emergency planners should have anticipated a tsunami wave of this magnitude and sited the diesel generators at a higher elevation or constructed a higher sea wall to block the tsunami wave (or both) is being debated. Should a reactor even be built by the ocean? Many are on rivers, where waves are not an issue. However, as we've noted, the ocean provided a relatively safe place for the released radioactivity to fall, compared to what happened after the Chernobyl accident.

Adding to the tragedy, in July 2012 the Fukushima Nuclear Accident Independent Investigation Commission concluded that the disaster likely could have been prevented had proper safety guidelines and government regulations been followed. The report placed responsibility for the disaster directly on collusion between the Tokyo Electric Power Company (TEPCO), the government, and the regulators. They "betrayed the nation's right to safety from nuclear accidents," the commission concluded, and added that TEPCO "manipulated its cozy relationship with regulators to take the teeth out of regulations."

"It was a profoundly man-made disaster—that could and should have been foreseen and prevented," the commission's chairman, Kiyoshi Kurokawa, is quoted as saying in the report's introduction. "And its effects could have been mitigated by a more effective human response." Not everyone agrees with these conclusions.

The potential benefits of nuclear power are undermined, and public safety is compromised, without rigorous adherence to the guidelines for operational safety. This is equally true for the production of electricity from any source. However, the long-term consequences and costs from a nuclear power facility accident are higher than for any other source of electrical generation. Or so they seem: it is easier to define a discrete area of radioactive contamination than to quantify the diffuse economic and political effects of reliance on foreign sources of oil or the effects of hydrocarbon use on global climate change.

CHAPTER 4

RADIATION AND CANCER

HOW DOES RADIATION CAUSE CANCER?

The explanation of why radiation causes cancer does not rest entirely on a predictable event. A genetic predisposition may play some role but there is also a large contribution from unpredictable events referred to as randomness. When ionizing radiations pass through cells, they change a molecule or an atom into an ion. Some of these occur in the nuclei of cells, and some within the cells' DNA. Most of the DNA does not encode genes. Rather, it is a collection of what until very recently were thought to be nonfunctioning genes (sometimes called "junk DNA") that just provides spacing, or repeats genetic sequences, or is detritus picked up over evolution from viruses, microbes, and other things. Now, however, it is believed that there are at least 4 million gene switches in it that can affect how cells and other tissue act. Of this mass of DNA, less than 1 percent is genes that encode proteins.

Think of DNA as a string on which genes sit, much like pearls in a necklace. If ionization occurs somewhere along a part of the string that has no pearl (gene), it may have no important consequence (except when the string is broken). Even if an ionization occurs within the part of the necklace where there is a pearl or gene, it may not be important, because genes are made up of important and less important parts called exons and introns. Another alternative is that the ionization damages a critical part of a gene so catastrophically that the cell dies. This is unfortunate for the cell but fortunate for us: a dead cell cannot cause cancer. However, very rarely and randomly, radiations passing through a cell may cause a nonfatal ionization in a critical part of an important gene or regulatory sequence that, alone or in combination with other genetic or epigenetic events, leads to cancer. Such an event where the DNA of the cell is altered is called a mutation. Mutations in DNA underly all or almost all cases of cancer. Radiation is not the only cause of mutations in DNA, but it is an important cause.

Some sophisticated biology and some big numbers are needed to fully explain how cancer is caused. The probability that a substance will cause cancer is rather like a game of unimaginably small darts in which only a bull's-eye counts in the scoring. One billion cells weigh 1 gram (0.03 ounces), so a kilogram (2.2 pounds) contains 1,000 billion cells—a trillion. The average 50-year-old American or European male weighs about 200 pounds (90 kilograms) and therefore has about 90 trillion cells in his body. Almost every cell contains 10 feet of coiled DNA that is not visible, even under a conventional microscope. These 10 feet are composed of about 23,000 genes.

When someone is exposed to radiation, the X-rays, gamma rays, electrons, alpha particles, or protons go through those 10 feet of coiled DNA. But these waves and particles are very small (actually, some have no size) compared to the size of the nucleus of a cell where the DNA is. A neutron measures about 1 thirty-thousand-billionth of a foot. A dart tip is about 1 thirty-thousandth of a foot. A baseball representing a cell the corresponding size to the dart in our analogy would be 3 trillion times bigger than a regular baseball. Now imagine this super baseball filled with a length of tightly coiled garden hose. The likelihood of hitting a specific 1-foot "gene" on that hose is minuscule. The dart of radiation has to score a direct hit in a place that will cause a specific change to cause cancer. Yet it happens often enough to kill millions of people a year.

POLONIUM-210, RADON-222, AND CANCER

One way smoking cigarettes causes cancer is by delivering radioactive materials into the lungs. Smoking cigarettes puts the smoker at risk for five of the most common causes of death worldwide, including heart and vascular diseases, cancer, emphysema, and pneumonia. Worse yet, sudden infant death syndrome and premature births are more common in the children of smokers. There are 5.4 million tobacco-related deaths worldwide each year, 1.3 million of them from lung cancer. Approximately 443,000 Americans die each year from tobacco-related illnesses; about 50,000 of them are nonsmokers killed by the effects of secondhand smoke. (The number of nonsmokers killed worldwide is about 600,000.)

On average, smokers live about thirteen years less than comparable nonsmokers. Eight out of every 10 lung cancer deaths are caused by smoking. Yet 130 years ago, a physician who discovered lung cancer in someone would have found the condition so rare that it would have been worth reporting in the medical literature; by 1889 only 140 cases of lung cancer were reported worldwide. The first edition of the widely used *Merck Manual,* published that year, listed cigarette smoking as a possible treatment for bronchitis and asthma. It took until 1912 for someone to suggest that smoking caused lung cancer.

What happened? According to the World Health Organization's International Agency for Research on Cancer, tobacco smoke is approximately 5 percent tar and 95 percent gas, containing more than 4,000 chemically distinct compounds. At least 60 of these are known to cause cancer in animals. Eleven cause cancer in humans, including arsenic, benzene, radium-222, and thorium. In a word, however, what made smoking even unhealthier is not what is in the tobacco but what is on it: fertilizer.

Fertilizers used in growing tobacco and many other plants are rich in chemicals, including radium-226 and its decay products, including polonium-210. Anything grown where there is uranium-238 in the ground or with the aid of these fertilizers picks up radionuclides from uranium-238 decay products, including radium-226, which gathers in their leaves, as with spinach or broccoli. This is not necessarily dangerous. About three-fourths of the polonium-210 that enters our bodies comes from our food, and 50 to 90 percent of it is excreted in our sweat and feces. The remainder, however, stays in the body and circulates via the blood throughout the body; cigarette smokers have 30 percent more polonium-210 in their blood than nonsmokers.

When someone lights a cigarette and draws in the smoke, the temperature of the tobacco in the middle of the red ember is about 1,500 degrees Fahrenheit (800 to 900 degrees Celsius). This aerosolizes the polonium-210, which is then inhaled directly into the bronchial tree and lungs. Filters are of little help; they reduce inhaled polonium-210 levels by less than 5 percent. Polonium-210 on the leaves of, say, spinach or broccoli is not dangerous, unless for some reason you choose to smoke them and thus aerosolize it. But smoking a cigarette is, in some regards, like intentionally inhaling a small nuclear weapon into your lungs.

Cigarette manufacturers have known about the presence of polonium-210 in tobacco since the 1960s.

Polonium-210 in the wrong part of a body can be deadly in even the smallest amount. (It is 250,000 times more toxic than cyanide.) A lethal dose of polonium-210 weighs 0.1 microgram—a tenth of a millionth of a gram, or less than a millionth of the weight of a snowflake. It has been used at least once as a murder weapon. In November 2006, Alexander Litvinenko, a former officer in the Federal Security Service of the Russian Federation (FSB, the successsor to the KGB), then living in London, who claimed in two books that Vladimir Putin's rise to power was organized by the FSB, fell mysteriously ill and died three weeks later. An autopsy showed acute radiation poisoning from polonium-210, which was traced to a Russian nuclear power facility, and thence to two FSB agents who handled it before surreptitiously slipping it into Litvinenko's food at a London sushi restaurant. This in some ways was an early case of nuclear terrorism or at least nuclear assassination. It certainly is the first documented use of polonium-210 to kill someone. Doctors failed to recognize what was going on until after Litvinenko died because—unlike most powerful radionuclides, which emit easily detectable gamma radiations—polonium-210 emits alpha particles that cannot penetrate human skin or even a piece of paper. Consequently, radiation detectors outside his body showed normal readings. But once introduced into a body, alpha particles cause considerable cell destruction and death. There are recent unsubstantiated claims that Yassir Arafat may also have died of polonium-210 poisoning.

If polonium-210 is that dangerous, and smoking causes lung cancer, why do only some people who smoke develop lung cancer? There is an obvious randomness to such a rare event as a DNA mutation, although genetic factors may play a role in some people. Nevertheless, the probability of such an event is clearly related to the amount of radiation passing through a lung cell. This explains why there is a correlation between how much someone smokes (increasing amounts of radiation from polonium-210) and lung cancer. It also explains why not everyone who smokes develops lung cancer: randomness. Still, a male smoker has a twentyfold greater chance of getting lung cancer than a comparable nonsmoker. The risk is about thirteenfold greater for female smokers than nonsmokers. The more one smokes, the more likely one is to get lung cancer. Period. Keep in mind that polonium-210 is not the only reason smokers get lung cancer, but it is an important one. These same considerations probably also apply to lung cancer caused by exposure to radon-222.

Which brings us back to our darts analogy. The more rays or particles that go through a cell, the more likely one is to cause a mutation capable of causing cancer. Like all cancers, it needs to happen only once to be deadly, and it happens often enough worldwide for lung cancer to kill more than 1 million people annually.

Estimated Life Expectancy Lost

Cigarette smoking	6 years
Overweight	2 years
Alcohol	1 year
Accidents	207 days
Natural hazards	7 days
Radiation (3 mSv)	15 days

1 in 100,000 Chance of Death

Smoking	14 cigarettes
Eating	25 cups of peanut butter
Spending	20 days in New York
Driving	375 miles in a car
Flying	25,000 miles in a plane
Canoeing	for 1 hour
Radiation exposure	0.2 mSv

STRONTIUM-90 AND SARCOMAS

It is human nature to trust our senses. If we put a hand too close to a fire, we automatically jerk it away when it becomes too hot. If we look down from the edge of a cliff, we step back. One reason the topic of radiation is so disturbing and causes disproportionate fear is that we can't trust our senses to detect it. We can't touch it or feel it or even see it (we can't detect most of the electromagnetic spectrum—all we can see is the tiny part that constitutes visible light); and we can be exposed to a huge dose without knowing it. Moreover, many people view nuclear power as being riskier than fossil fuels, much as they fear dying in a plane crash more than in a car crash. The annual risk of being killed in an airplane accident for the average American is 1 in 2 million, whereas the risk of death in a motor vehicle accident is about 1 in 8,000. In absolute terms this translates to average deaths of about 140 versus 37,000 each year. Nevertheless more people seem to fear dying in an airplane crash much more than they fear dying on the road, for two reasons. One is predictability: deaths from air crashes are mostly unpredictable, whereas deaths in car crashes, though random, occur at a reasonably predictable rate each year. People do not like unpredictability (although very few spend time predicting their death). The second reason has to do with the number of people killed in any single accident. When a plane crashes, 300 people may die instantaneously. But in a fatal car crash usually one or two people die. The idea of a large number of deaths in one instance is unsettling. The net result is that many people are afraid to fly but see no danger in driving to the airport. Yet by the time they get to the airport, most of their risk is over.

We can use this example to look at the issue of using nuclear energy versus fossil fuels to generate electricity. There has been only one instance of unpredicted radiation-related deaths attributable to nuclear power (Chernobyl), but this is scarier to most people than the large but predictable number of deaths attributable to producing electricity from fossil fuels. If you stand unwittingly in front of an exposed radiation source, you wouldn't be aware of being exposed unless you received a dose of 1,000 to 2,000 mSv—around one-third of a dose that could kill you without medical intervention. When we are dealing with small doses of radiation, it is usually impossible for a person to know they are being exposed. For instance, as you now know, the average annual total radiation dose for Americans is 6.2 mSv. If instead you received 100 times more one year, 620 mSv, you would never know it. And even if you received that dose instantaneously, you would not know it. The point is, we usually cannot discern how much radiation we are exposed to. The only visible part of the body affected by a high dose of radiation is the skin, which turns red for many reasons—blushing, for example, or too much hot sauce—other than exposure to the sun. This makes skin changes an unreliable measure of radiation exposure.

An interesting exception to the difficulty of measuring radiation exposure came in 1959. St. Louis physician Dr. Louise Reiss, convinced that radioactive fallout from atmospheric nuclear testing was entering America's food supply and ending up in people's bones and teeth, thought of a simple and noninvasive way to test that hypothesis: study the amount of radioactivity in the baby teeth that children normally shed. Like scientific tooth fairies, Reiss and her colleagues visited hundreds of St. Louis–area schools, scout groups, YMCAs, churches, and synagogues, asking for baby teeth and distributing kits for children to mail them in. In return, children received a button that read, "I gave my tooth to science." So many thousands of teeth arrived at the Reiss home that often as many as thirty women volunteers sorted them out on card tables. The first data on the accumulation of strontium-90 in children's teeth was published in the journal *Science* in November 1961.

For more than a dozen years Reiss, her husband, Eric (also a physician), and other scientists at Washington University and St. Louis University collected about 320,000 baby teeth. Reiss and her colleagues showed that children born in St. Louis in 1963 had 50 times more strontium-90 in their teeth than those born in 1950, when only a few atmospheric nuclear tests had been carried out. But how dangerous was this level?

Strontium in its natural state is a soft, shiny, silvery-gray metal that quickly turns yellowish after exposure to air. It was discovered in lead mines in the Scottish village of Strontian in 1787 and is the fourteenth most common element on Earth. Strontium produces the red sparkle in fireworks and signal flares, and in some parts of the world it is considered healthy to add it to food: in China, for instance, bottlers of spring water with a high amount of strontium advertise the content on the label. Strontium-88 is the element's stable, nonradioactive form. But when an atom of uranium-235 fissions in a nuclear weapon or in a nuclear reactor, strontium-90, a radioactive form of strontium-88, is created (along with about two hundred other fission products).

Strontium-90 has benefits as well as dangers. It is incorporated in a stent placed in a blood vessel in the heart to retard growth of the surrounding tissue that might enter and occlude the stent. Strontium-90 is also used to treat pterygium, a noncancerous overgrowth of the tissue (conjunctiva) that overlies the white part of the eye (sclera); it may be caused by UV radiations. Because strontium-90 produces a great deal of heat when it decays, that heat can be turned into electricity for a long-term power source in a nuclear reactor; or in an isolated area, such as a lighthouse in Antarctica, where months pass without sunshine, to produce electricity from solar energy; or for Coast Guard buoys in the ocean. Strontium-90 is also used to power spacecraft whose voyages last beyond the life of a solar cell or that travel far from the Sun and thus have no way to gather solar power, like the 2008 Cassini Solstice Mission sent to explore Saturn and its environs that will continue beyond 2017.

Some people are concerned about sending a radioactive device into space—which, of course, is already radioactive and the ultimate source of all naturally occurring radiation. They also question what would happen if a rocket with strontium-90 aboard were to explode during launch. That small amount of onboard strontium-90 is unimportant in the context of radionuclides released by previous atmospheric nuclear weapons testing. Also, as we will discuss in chapter 8, the oceans contain large amounts of naturally occurring and man-made radioactivity, and the small amount of strontium-90 deposited in the water would be inconsequen-

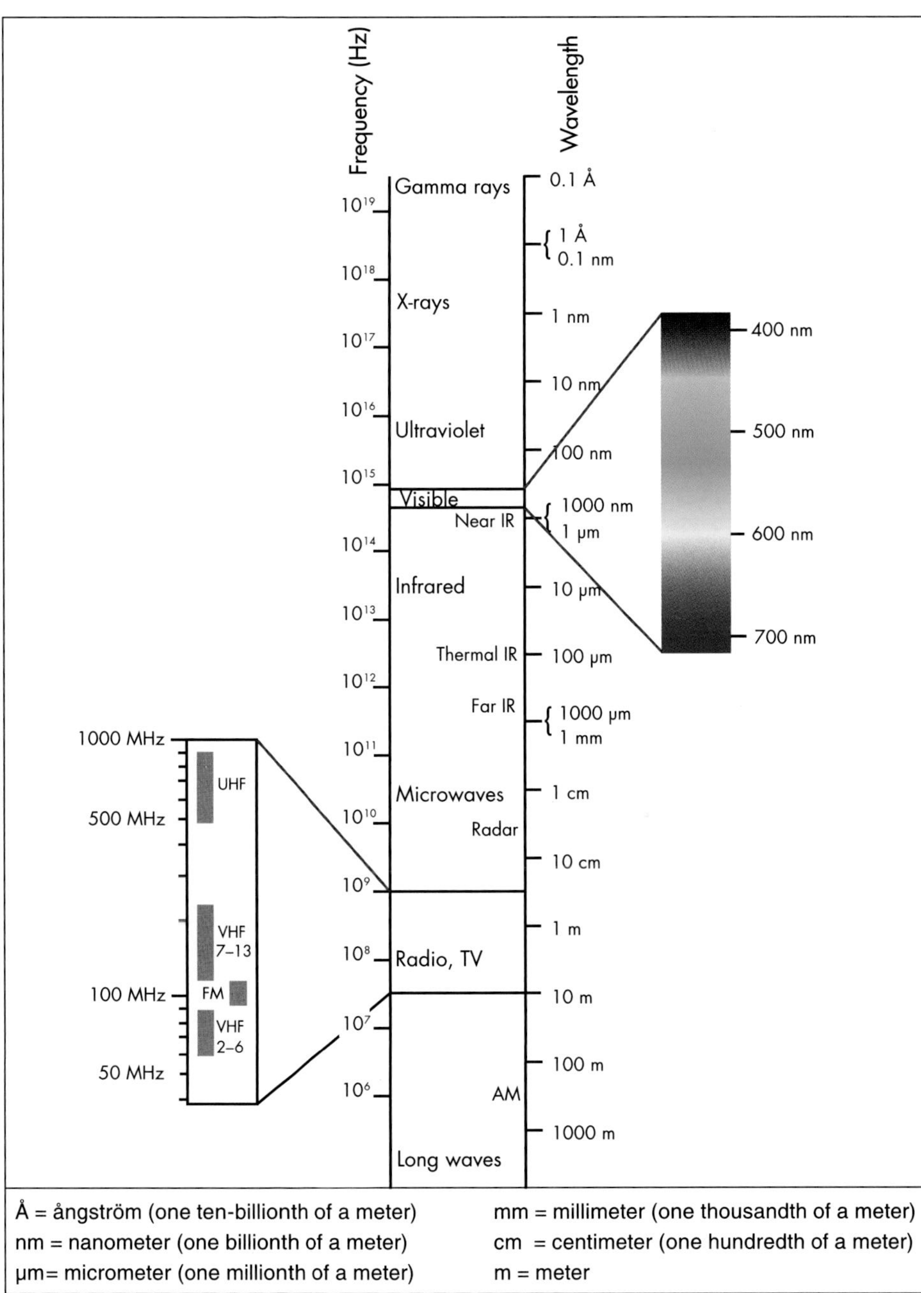

Visible light is but a tiny part of the energy that affects us. The electromagnetic rays beyond our visibility toward the top grow increasingly dangerous to humans as their wavelengths grow shorter.

SOURCES OF RADIATION EXPOSURE IN THE UNITED STATES

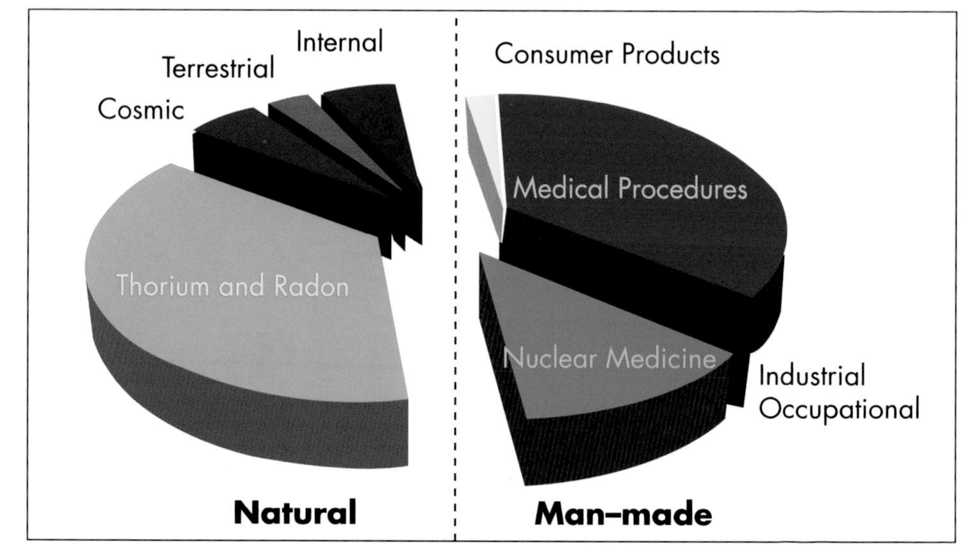

Radiation received by people in the United States is almost evenly divided between man-made and natural sources. People living in most other parts of the world receive less radiation from medical procedures and nuclear medicine.

TERRESTRIAL GAMMA-RAY EXPOSURE AT 1 METER ABOVE GROUND

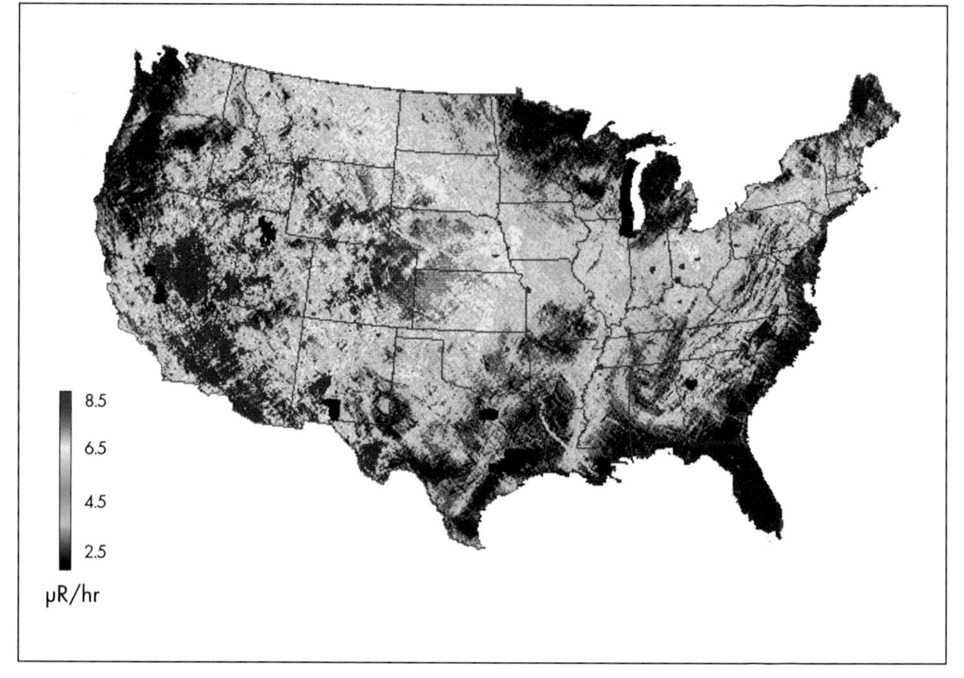

The annual terrestial radiation we receive is affected by where we live.

CHERNOBYL, DISPERSAL OF CESIUM-137

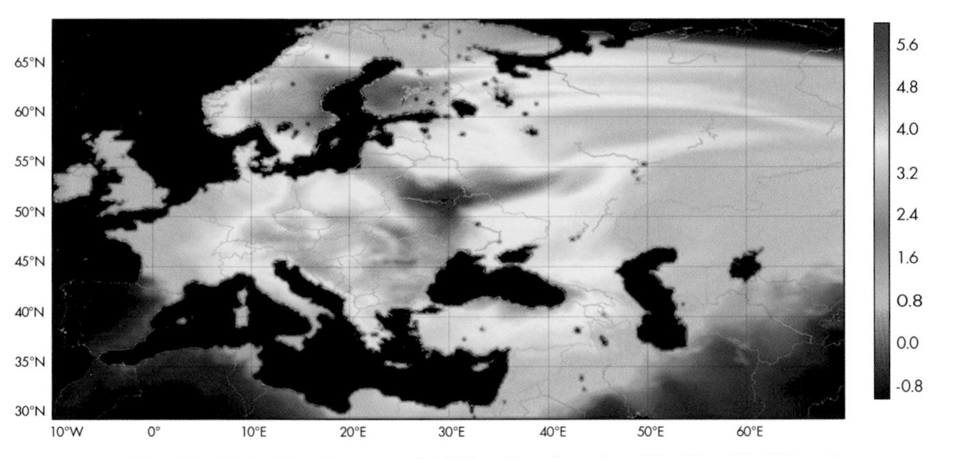

FUKUSHIMA DAIICHI, DISPERSAL OF CESIUM-137

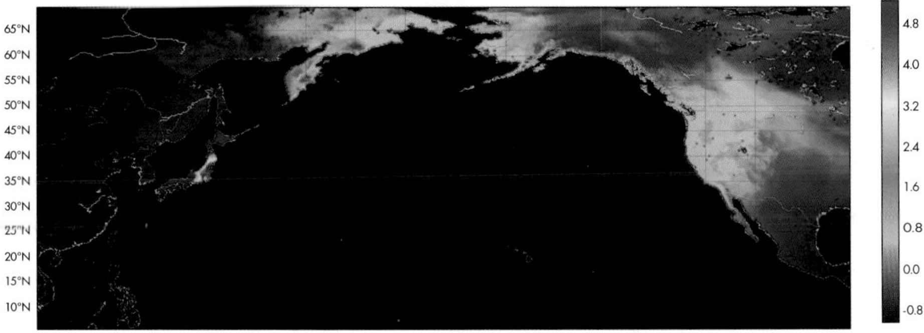

Cesium-137 carried by the wind after nuclear accidents in Chernobyl and Fukushima. The red, orange, and yellow show a much greater release of radiation and a much wider dispersal from Chernobyl than from Fukushima. (These illustrations are results of numerical modeling, and quantitative errors made in the field of radionuclides modeling are usually large. They do, however, show the difference between the two accidents.)

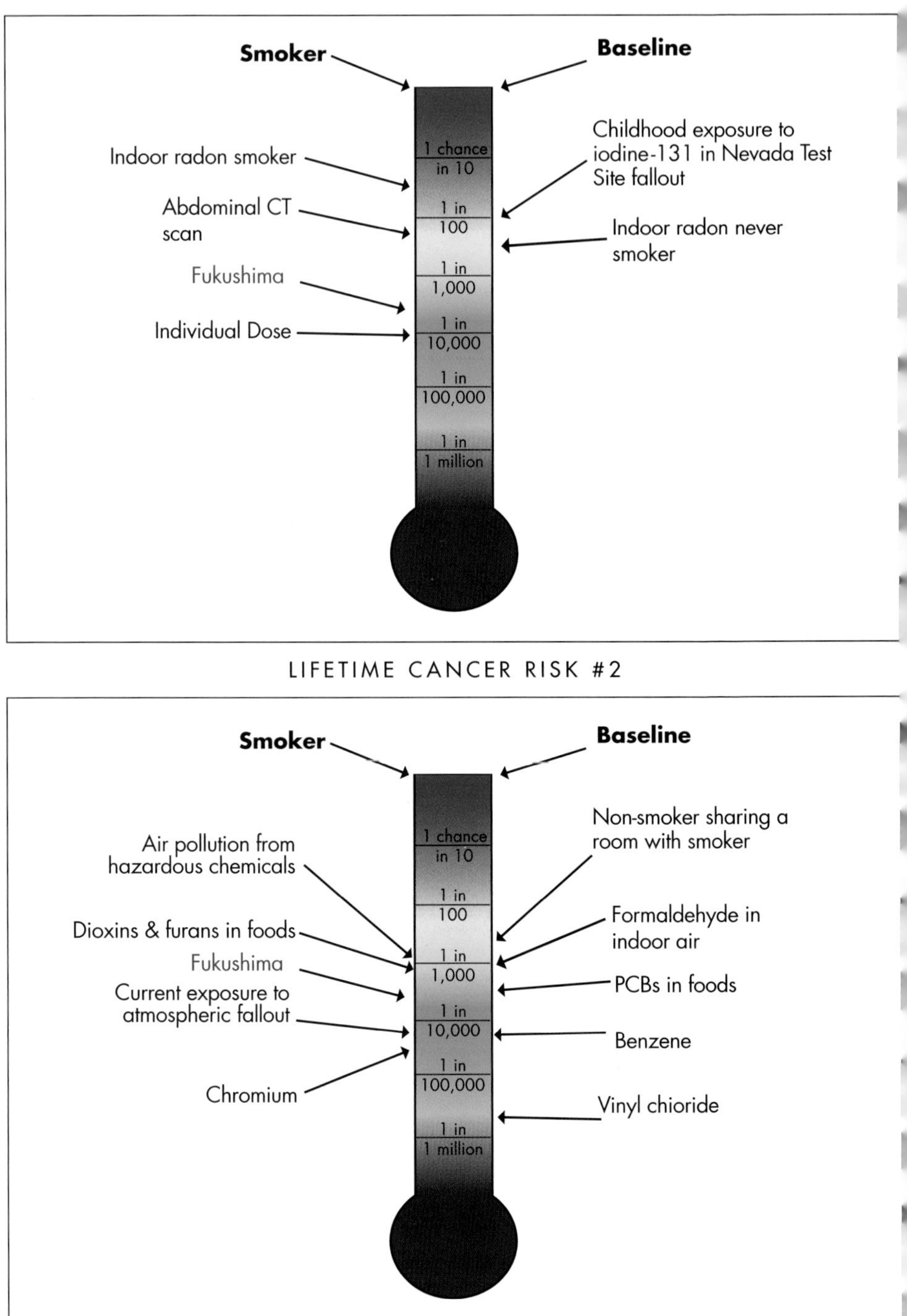

Two ways of looking at the probable cancer risk associated with radiation released from Fukushima Daiichi compared to other health hazards.

tial. As for reentry, the rocket would likely burn up or fall into the tremendous volume of the oceans and disappear without a trace.

Strontium-90 is absorbed into the body from water and dairy products. Because the body responds to it as if it were calcium, the 70 percent or so that is not excreted in urine or feces settles for decades in the teeth and bones. It can cause cancer in or around the bones.

The half-life of strontium-90 is nearly 30 years, so the strontium-90 we ingest has the potential to remain in us (and our remains) for around 300 years, as we've mentioned. Worse, the biological half-life of strontium-90 is 49 years. If you ingest a dose of strontium-90 at age 1 and you die at age 70, you will still have in your body roughly a quarter of what you ate.

In 1963 Dr. Eric Reiss testified before a U.S. Senate committee in support of the ratification of a treaty between the Soviet Union, Great Britain, and the United States that would halt atmospheric testing of nuclear weapons. Between 1945 and 1963, the United States conducted 206 atmospheric weapons tests, almost evenly divided between the Nevada Test Site and the Marshall Islands in the Pacific Ocean; another 216 were carried out by the Soviet Union; and Great Britain performed 7 in Western Australia. Evidence of strontium-90's potentially deleterious impact on children's health was so compelling to President John F. Kennedy that he signed the Partial Test Ban Treaty in August 1963. (France, which exploded 49 nuclear devices in the atmosphere in Algeria and French Polynesia, continued to conduct atmospheric nuclear tests until 1974. China performed its last atmospheric nuclear test in 1980.)

But what did this accumulation of strontium-90 actually mean for the children of St. Louis and all those people who were hundreds or thousands of miles downwind (that is, generally to the east) of the Nevada Test Site? The precise outcome is still being debated, but it seems that the effect was small. The more significant result of the Reiss data is that it helped halt atmospheric nuclear testing.

Sarcomas—cancers of bone, cartilage, muscle, and joints—are radiogenic cancers, though radiation is only one of several possible causes. Radiogenic sarcomas are caused by certain bone-seeking radionuclides like strontium-90 that, if inhaled or ingested, settle in bone. External high-dose radiation, usually given to treat another cancer can also cause sarcomas. If someone is treated for a cancer with radiation and 10 or 20 years later develops a sarcoma in the radiation field, the radiation likely induced the sarcoma.

The safest assumption is that all types of cancer can be caused or contributed to by radiation. This is based on the notion that the inability to show an effect in epidemiological studies does not mean the effect does not exist. Different types of cancer, however, have different sensitivities to induction by ionizing radiations. Leukemias, for example, are much more radiogenic than cancers of the bone. Most other cancers fall between these extremes.

In sum and in general, the types of cancer increased by radiation are those that occur normally within a population. As we mentioned in chapter 2, chronic lymphocytic leukemia is rare in Japanese, and its incidence was not increased in the atomic bomb survivors, whereas other types that occur normally in Japanese were increased. Thyroid cancer, which normally is more common in females than males, was more common in girls who were exposed to radiation from the Chernobyl accident than it was in boys.

Most of the cases of radiation-induced breast cancer in the atomic bomb survivors were detected in women who were young girls at the time of radiation exposure. The same is true for young persons who were exposed to radioactive iodine-131 after the Chernobyl accident. Some estimates indicate newborns are ten to thirty times more susceptible to cancer induction from iodine-131 than adults.

It is important to remember that these data refer to ionizing radiations. Nonionizing radiations, like those associated with microwaves and cell phones, are not convincingly associated with an increased cancer risk. Moreover, there is no proven biological mechanism whereby nonionizing radiation might cause cancer. Substantial debate rages about whether excess cell phone use causes brain cancers. However, there has been no increase in brain cancer deaths over the past 20 years despite an extraordinary increase in cell phone use. Although some scientists argue it may be too early to detect such an effect, this seems unlikely. (For more detail on this, see page 215.)

SKIN CANCERS AND UV RADIATION

Melanoma is less common than other skin cancers—less than 5 percent of all cases—but it accounts for 70 percent of skin cancer deaths. It is a cancer of the melanocytes, cells that produce the dark pigment melanin, which colors the skin. Although melanocytes are found in the skin, they are really a form of nerve cell. Melanocytes are also found in the eye and near the covering of the brain, the heart, the bone, the mucous membranes, and the gastrointestinal tract. Consequently, melanomas can occur at sites other than the skin. A suntan is the result of UVB radiations stimulating melanocytes to produce melanin, in an attempt to protect the deeper layers of the skin from UVB-induced DNA damage from the Sun (or sometimes a tanning lamp). After melanocytes make melanin, they pass the pigment to skin cells called keratinocytes. Keratinocytes have a short lifespan, about four days, whereas melanocytes live many years, perhaps for your entire life. When exposure to sunlight is withdrawn, a person loses their tan fairly rapidly because the melanin-containing keratinocytes are shed and melanin production by the remaining melanocytes decreases. (This is why you don't want to have a facial scrub after a tanning session at a beauty salon. In fact, you might consider skipping both.)

As with all healthy cells, melanocyte growth is carefully regulated so their numbers remain relatively constant throughout a person's lifetime. However, if the DNA in a melanocyte is damaged by UVB radiation or by an environmental or genetic factor or both, growth of the affected cell may become unregulated. This unregulated growth is what constitutes a cancer, in this case a melanoma. Melanoma is more prevalent in women than men before age 40 and in men after age 40. Melanoma is also more prevalent in persons of European descent, especially Celtic peoples, and is 10 to 20 times more prevalent in whites than in blacks living in the same area and exposed to the same amount of UVB, because blacks have more melanin to protect their melanocytes from UVB-induced mutations. When blacks get melanoma, it is often in lighter skin areas like the palms and soles. Melanoma is uncommon in Asians and in Hispanics, including Hispanics of European descent, for unknown reasons. These differences are interesting as all humans have about the same number of melanocytes in their skin but differ in how much melanin they produce. Because the role of melanin is to absorb UVB radiations and protect against mutations in DNA, this is good evidence that UVB causes melanoma. However, this notion does not explain why comparably pigmented older males have much higher melanoma rates than females. (Perhaps they are more likely to work out of doors and have more UVB exposure.) Pigmentation also does not entirely explain why Asians have low melanoma rates similar to blacks' and Hispanics' rather than the higher rates of whites.

About 200,000 cases of melanoma are diagnosed annually worldwide, predominantly in sunny climates—Australia, Latin America, New Zealand, and North America. The World Health Organization estimates that each year there are about 65,000 melanoma-related deaths worldwide.

In the past several decades, melanoma cases have increased at an alarming rate, as much as 800 percent among young women and 400 percent amoung young men. Some scientists speculate that although it was once not considered fashionable to have a suntan, that changed in the twentieth century, and now there are people in their fifties, sixties, and seventies who were widely exposed to sunshine when younger. One theory is Coco Chanel started the tanning craze by showing models with less than lily-white skin. Whatever the cause, a suntan, which had always been the mark of an agricultural or other outdoors worker, suddenly was fashionable in some societies. (This is not so in most Asian countries. Women in Tokyo and Beijing are often seen carrying umbrellas and wearing gloves on clear days in midsummer.)

People in sunny climes who slather on sunblock usually think they are protecting themselves against melanoma. Unless they have been using it since birth, however, they are not doing much in the way of prevention, although they are wisely protecting themselves against other skin cancers and skin damage. Two studies—one in 1985, another in 1992—show that a person's likelihood of getting melanoma is generally determined at a very early age. People who move from an area with a low incidence of melanoma to an area with a high rate—sunnier countries—generally have lower rates of melanoma than those born in the sunnier climate. People born and raised in Britain who moved to Australia and New Zealand (where the population is primarily of British origin, so similar genetic traits can be compared) have a mortality rate about half that of people born in the sunnier climes. The exception is those who were 10 or younger when they arrived; their instances of melanoma are equal to those born in the sunnier countries. They are equally as susceptible to sun damage that leads to melanoma in later life.

Americans who grew up in the 1950s and 1960s thought the best way to achieve a good suntan was to coat their skin with baby oil, thus basting themselves and magnifying the effects of (and damage caused by) UV rays. Since then it has been proved that the best protection against UV rays is to keep direct sunlight from your skin by wearing long-sleeved clothes and a broad-brimmed hat. Sunblocks and sunscreens are effective if used properly, but a 2009 poll by *Consumer Reports* showed that 31 percent of Americans never use sunscreen, and only about 50 percent use it regularly. A 2012 survey by the American Academy of Pediatrics showed that only 25 percent of children use sunscreen regularly. Sunblocks, such as zinc oxide and titanium dioxide, physically block UVA and UVB rays, but they are clearly visible, thick, greasy, and cosmetically unacceptable for most people. Sunscreens chemically filter and reduce UV penetration of the skin; they are labeled with SPF (Sun Protection Factor) numbers that range from 2 to 50 and higher. SPF 15 means you can be in the sun 15 times longer before burning than you can if you apply none, but these numbers are not proportional. A lotion with SPF 2 absorbs 50 percent of UV radiation; SPF 15 absorbs 93 percent; SPF 34 absorbs 97 percent.

Ultraviolet radiations are so called because they are proximal to the color the human eye perceives as violet. UV radiations mostly emanate from the Sun, but they have some man-made sources, like tanning lamps and arc welding devices. UV radiations have a shorter wavelength (are more energetic) than visible light but a longer wavelength (less energetic) than X-rays. Most UV radiations lack sufficient energy to cause ionizations and are therefore classified as nonionizing. This contrasts with the forms of radiation we have discussed up to this point. However, some UV radiations have sufficient energy to alter chemical bonds, the forces that hold atoms together. This energy is the basis of their cancer-causing ability. (A caveat: rare high-frequency UV radiations can cause ionizations under appropriate circumstances.)

There are three forms of ultraviolet radiation: UVA, UVB, and UVC. UVA is the most ubiquitous. It can penetrate the skin and alter skin cells and the collagen that is their supporting structure. This results in the appearance of aging and in cancers. UVB waves have more energy to alter chemical bonds, but most are absorbed as they pass through Earth's atmosphere. They are closely correlated with the risk of developing melanoma. UVC waves are the most energetic. Fortunately, almost all UVC waves are absorbed in the atmosphere.

The Sun's radiations are measured by their wavelength, and, as with all radiations, the shorter the wavelength, the greater the capacity to cause biological and chemical changes. The colors of the light we are able to see—in order of increasing energy: red, orange, yellow, green, blue, violet—range in wavelength from more than 700 nanometers (nm; 1 nanometer is one-billionth of a meter) to about 400 nm.

Infrared radiation (that is, radiation below the red end

of the visible spectrum with the longer, less powerful, and therefore less harmful wavelengths that range from just over 700 to about 995 nm) was discovered in 1800 by the German-born English astronomer and composer Sir Frederick William Herschel (1738–1822), who gained greater fame for the discovery of Uranus and its two major moons, Oberon and Titania, as well as two moons of Saturn. Herschel's infrared radiation was referred to as "heat rays." The next year Johann Wilhelm Ritter (1776–1810), a German scientist and philosopher who had an abiding interest in the polar forces of nature, looked for the opposite, cooler end of Herschel's discovery and detected radiation above the shorter, violet end of the visible spectrum: ultraviolet radiation (100 to 400 nm).

Types of skin cancer other than melanoma include basal cell and squamous cell carcinomas. If you have a fair complexion and light eyes, and if you freckle and sunburn easily, you are at greater risk for these cancers than people without those characteristics. Skin cancers generally appear where there has been the most exposure to sunlight. About 95 percent of the sun's rays are UVA radiation, which are the dominant tanning ray because of their ubiquity. UVB rays are more intense but less common than UVA; they are the major cause of sunburn, and reddening of the skin, and of course melanoma. They are more potent at higher altitudes because passing through the atmosphere moderates the intensity of UVB rays. They are also more intense on water, ice, and snow, which reflect back about 80 percent of UVB rays. UVA has long been known to play a major part in aging and wrinkling the skin, but until around 1990 scientists thought UVA did little damage to the outer layer of skin, the epidermis, where most skin cancers occur. Now, however, considerable evidence shows that, along with other environmental and genetic factors, UVA contributes to skin cancer and may directly cause the damage that begins a basal cell or squamous cell carcinoma. Tanning booths give off about 12 times as much UVA as the Sun. They have been declared a carcinogen by several health authorities, including the World Health Organization.

Solar rays below about 290 nm virtually never reach the earth's surface because they are absorbed by the atmosphere, but a small amount of UVB (290 to 320 nm) does. Even though UVB represents less than 1 percent of the total energy released by the Sun, it is considered 3 to 4 times more likely than UVA to cause sunburn in humans and skin cancer in lab animals. Among the benefits of UVB is that it helps the body synthesize vitamin D, necessary for the regulation of many body systems, especially bone remodeling as old bone is replaced by new. It also prevents rickets, a softening of the bones caused by a deficiency of vitamin D, calcium, or phosphate. Rickets now appears mostly in people with poor nutrition in underdeveloped countries and in dark-skinned people in urban environments who don't spend much time outside.

Basal cell carcinoma is the most common cancer in the United States, and the most common in persons of European descent worldwide. New cases are growing at a rate of about 10 percent a year. It can occur in young people but is most common in those older than 40. It starts in the epidermis and occurs most often on whatever skin—including the scalp—is regularly exposed to sunlight or other UV radiation. Typically it spreads slowly and does not spread to distant sites. If not discovered and treated by freezing or excision, it eventually can grow rapidly and may require radiation treatment. If a basal cell carcinoma is left to grow too long, the necessary surgery can be disfiguring because of the amount of tissue that has to be removed. The World Health Organization estimates that about 2.8 million cases are diagnosed each year in the United States (and about 10 million cases worldwide) but account for only 1 in 1,000 cancer deaths.

Squamous cell cancer may occur in normal skin or in skin that has been injured or inflamed. Most skin cancers occur on skin that is regularly exposed to sunlight or other UV radiation. The earliest form of squamous cell skin cancer is called Bowen disease (or squamous cell in situ). This type does not spread to nearby tissues. Squamous cell cancer spreads more rapidly than basal cell but is still slow-growing. It accounts for less than 1 percent of annual cancer deaths in the United States.

Actinic (*aktis* means "ray" in Greek) keratosis produces dry, flat, crusty patches of skin in sites of substantial sun exposure. It is a precancerous skin lesion that rarely becomes a squamous cell cancer.

Although for most of us the greatest worry about UV rays is sunburn, children with xeroderma pigmentosum (XP), an inherited genetic disorder in which the body is unable to repair damage done by UV light, are really in trouble. XP affects about 1 in a million Americans. (Japanese are about 6 times more likely to have it.) People with XP have a risk of skin cancer about 5,000 times greater than normal. All forms of life develop protective mechanisms to adapt to the environment. If during evolution the human body had not developed an enzyme pathway to repair UV light–induced DNA damage (cross-links), humans would not be able to spend *any* time in the sun. We live as we do because we have developed a way to protect against getting skin cancer, at least for the most part.

One reason for this protection is that the ozone layer absorbs most of the UV rays in the stratosphere 15 to

80 miles above the Earth's surface. But that protective layer is being steadily depleted by human use of nitrogen fertilizers, which emit nitrogen oxide gas, and by other man-made causes, including atmospheric nuclear weapons testing, as well as the burning of fossil fuels. In the mid-1970s it was discovered that chlorine atoms in chlorofluorocarbon compounds (CFCs), used primarily as refrigerants and propellants, absorb significant ozone. As the ozone layer decreases, the threat to humans, including more skin cancers, will increase because more UVB rays will pass through the stratosphere.

Although UVA rays may be a danger to most people, they are greatly beneficial to those with psoriasis, a malady of the skin sometimes associated with arthritis, where the skin becomes red and thick, with silvery patches that flake off like scales. It is not a cancer but is thought to be the result of the immune system's response to what seem to be new skin cells intruding in the body. Normally, skin cells grow deep in the skin and then rise to the surface in about a month. Psoriasis speeds up this process, causing dead cells to build up on the epidermis. A dye synthesized from psoralen (a photosensitizing chemical found in plants) is injected into a person with psoriasis and accumulates in the diseased skin. When the skin is exposed to UVA rays, the psoralen cross-links the DNA in the psoriasis cells and kills them.

In a sense, radiation has contributed to humans' having varied skin pigmentation. If we all share a common ancestry that began in Africa (and this is open to debate), why are there such differences in the color of people's skin? The genetic reason that people in equatorial places are darker skinned is that the climate supports the growth of fruits and berries, which, along with the abundant sunshine, provide a high amount of vitamin D. Studies show that incidence of skin cancer is inversely correlated with latitude. As humans migrated away from the equator to climates that were more suitable for growing grains, the only source of vitamin D became its synthesis from sunlight. The farther north or south one went, and the longer the periods of seasonally extended darkness, the smaller the amount of vitamin D that was available. As lighter skin absorbs more UVB than darker skin, losing skin color meant being able to capture more sunlight to make vitamin D. Natural selection favored lighter skin the nearer humans moved toward the poles, and darker skin where intense sunlight is a constant.

CHAPTER 5

GENETIC DISEASES, BIRTH DEFECTS, AND IRRADIATED FOOD

A deeply held fear of many people is that radiation damage to reproductive tissues (ovaries and testes) can be passed along to their children. As we mentioned, there were stories of three-headed cows and children with grotesque abnormalities after the Chernobyl and Three Mile Island nuclear power facility accidents. These abnormalities were attributed to radiation released by the accidents.

Before we dive into this complex and controversial issue, we need to define two terms that sometimes overlap and are used differently by different people: *genetic disease* and *birth defect.* A genetic disease or disorder is caused by an abnormality in a person's DNA inherited from one or both parents. In genetics, this abnormality is referred to as a *mutation*. The inherited nature of the mutation is key. For example, cancers are characterized by mutations in a person's DNA, but these mutations typically occur after birth and are present only in the cancer cells but not in the person's normal cells. In a person with a genetic disease, the mutation is usually present in all the cells of one (a dominant mutation) or both (a recessive mutation) parents. In some cases, however, it can arise because of a mutation only in a parent's germ cells rather than in the other cells of his or her body. Rarely, the mutation can occur in the affected person after the ovum is fertilized, in the early stages of embryonic development. These mutations will be present in all cells in the child's body and can be passed to the child's progeny when he or she reproduces (unless the mutation causes sterility). Some heritable abnormalities in humans are caused by a mechanism that affects the DNA indirectly. Such changes, called *epigenetic,* involve modification of the expression or function of the DNA, usually through nongenetic processes. Epigenetic abnormalities can cause or contribute to a genetic disease (and also to a birth defect).

Almost all cancers have DNA mutations that develop after birth and are not inherited. Consequently, cancer is not typically called a genetic disease, although one or many DNA mutations often underlie its development. There are, however, exceptions. Families with specific genetic mutations comprise 5 to 10 percent of new cancers. An example is heritable, or familial, breast cancer, associated with the BRCA1 and BRCA2 genetic mutations. Because these cancers are genetic and hereditary, the DNA mutation is present in every cell in a woman's body, not just breast cancer cells. Two common forms of colorectal cancer, Gardner and Lynch syndromes, are also inherited genetic diseases associated with cancer. On page 124 we discussed xeroderma pigmentosum, a genetic disorder with a high risk of skin cancers because of a defect in the excision and repair of mutations in DNA, caused by UV radiations. Genetic disorders are, by definition, present at birth, although they may have no readily detectable effect, such that the newborn appears normal. Because genetic diseases can be transmitted from generation to generation, they are of special interest in the process of evolution.

Birth defects are different from genetic diseases. They are abnormalities present at birth or detected soon thereafter. They may be genetic or not. For example, physical trauma during birth causes some birth defects but does not alter the child's DNA. Most birth defects are not heritable (cannot be transmitted to the child's children), although there are exceptions. DNA mutations occurring during meiosis (the process of cell division by which sperm and eggs are produced) are an example. Birth defects are caused by genetic or environmental mechanisms or even by both; genetic birth defects are caused by mutations in DNA. Usually the affected DNA is packaged in a chromosome in the nucleus of the cell, but sometimes DNA in the mitochondria (a subcellular structure inherited from the mother that contains some of the cell's DNA and is important in energy production) is mutated. Changes in the number of chromosomes are also important. For example, children with Down syndrome have an extra chromosome 21, whereas girls with Turner syndrome are missing one X-chromosome. Some birth defects are caused by exposure of the fetus to harmful agents like alcohol, tobacco, and specific types of antibiotics, anticancer drugs, psychotropic drugs, anticonvulsants, and hormones. Also, environmental agents like mercury and lead can cause birth defects. Substances that cause abnormalities in the embryo or fetus are referred to as teratogens; a well-known example is thalidomide. Other teratogens include viruses (like rubella) and microbes (like syphilis). Damage to a fetus resulting from a complication of pregnancy or delivery is also considered a birth defect. Some epigenetic causes of birth defects interact with genetic and nongenetic causes. High doses of ionizing radiations can also cause birth defects. But the causes of most birth defects are unknown. A search for genetic causes includes a detailed family history and often extensive DNA testing. Environmental causes may be revealed through the history or by excluding other explanations.

Birth defects are unfortunately rather common; about 3 percent of children have one or more at birth. In another 2 to 7 percent of children, an abnormality related to birth is not detected until months or a few years later. Adding these up, it is estimated 5 to 10 percent of live births are of children with birth defects. This high background rate makes it difficult or impossible to ascribe a birth defect to a specific cause unless there is a control group for comparison.

As we discussed on page 87, the Hiroshima and Nagasaki atomic bombings exposed about 3,000 pregnant women and their fetuses to ionizing radiations. In summary, 30 children exposed in utero in the second trimester were subsequently found to have a small head and mild to moderate mental retardation. These findings are usually attributed to adverse effects of radiation on the migration of neural cells from elsewhere in the body into the fetal brain, a process that occurs during the second trimester. The most affected children were those in the wombs of mothers who received about 1,000 mSv of radiation. (The radiation dose to the fetus is slightly lower than the dose to the mother because the fetus is shielded from some radiations by the uterus.) Children born to mothers exposed to less than 300 mSv had no detectable developmental abnormalities. Children with developmental delay from in utero radiation exposure often had a smaller than normal size head. Magnetic resonance imaging many years later showed that neurons in their brain migrated to the wrong place and led to abnormal brain architecture. Developmental abnormalities were not found in fetuses younger than 8 weeks and older than 25 weeks of gestation when their mothers were exposed to A-bomb radiations.

It was also important to determine whether genetic abnormalities were induced in children born to parents exposed to radiations from the A-bombs. Fortunately the answer seems to be no. Extensive analyses of genetic markers in the cells of these children show no increased rate of DNA mutations. Also, there is no increase in cancer or other causes of death in children of exposed versus nonexposed parents. Further studies of 77,000 children whose parents were exposed to radiations from the A-bombs showed no increase in birth defects, stillbirths, cancer, or other causes of death. A study of 12,000 children of A-bomb survivors when the children's average age had reached 50 years showed no increased risk of such common adult diseases as hypertension, diabetes, heart disease, and stroke.

So how should one interpret reports of increased birth defects following the Chernobyl and Three Mile Island accidents, and what should one expect after Fukushima? First, Chernobyl. Reports of birth defects in animals and humans fail to consider the high background rate of birth defects, 5 to 10 percent. Because of this high background rate, it is essential to compare rates of birth defects in a radiation-exposed population to that of an unexposed or less exposed population. No such study was done after the Chernobyl accident.

Another consideration in evaluating reports of increased birth defects is *ascertainment bias:* we tend to find more of something when we are looking for it, especially in a tragedy. Consider an everyday experience. A person buys a Volkswagen Jetta and then *suddenly* notices that everyone else seems to be driving a Volkswagen Jetta. We will find more birth defects in a population under scrutiny than we would otherwise find. This is simply human nature, which also seeks a cause, especially for untoward events. For a child born in Ukraine with a birth defect in 1987, it would be only normal (but scientifically incorrect) for the parents to ascribe this defect to radiations from Chernobyl. However, if we consider the high radiation dose correlated with birth defects in the A-bomb survivors (more than 300 mSv) and other data from radiation studies in animals and humans, it is very unlikely there could be birth defects directly attributable to radiation released by the Chernobyl accident. This is very good news for people living in Japan, where any impact of radiation released at Fukushima is extremely unlikely.

Other forms of radiation, however, can have an effect on a fetus. Between 1953 and 1956, Alice Stewart, a physician at the Royal Sheffield Hospital in England, detected an increase in leukemias in children exposed to X-rays in utero, usually to determine the fetal position in the mother's pelvis or to see whether the pregnancy was twins. Stewart's findings caused X-rays of pregnant women, infants, and young children to be curtailed worldwide. However, some of Stewart's conclusions are controversial.

Some have claimed to find genetic disorders, including cancer in the children of people occupationally exposed to radiation. The Sellafield nuclear power facility (formerly named Calder Hall), the first in Britain, opened in 1956; a study claimed an increased risk of leukemias and lymphomas in children of male nuclear workers. This finding has not been reproduced and is not widely accepted. It is difficult to imagine these data are correct because sperm have a very short life span. One of these short-lived sperm would have had to acquire one or more mutations from low-dose radiation over a short interval, then happen to be the one out of several million sperm to fertilize an egg resulting in a pregnancy; or else there would have to be a mutation in the germ cells in the testes that produce sperm. Both possibilities seem remote. As we discussed, children born to Japanese fathers exposed to the A-bombs, including those exposed to doses far higher than any possible exposure in a nuclear power facility, had no increased cancer risk. So the notion that an occupational exposure to radiation resulted in a mutation in a germ cell is unlikely. A third quite remote possibility is that low-dose radiation caused a mutation in a Sertoli cell, the cells that nurture the developing sperm. Because Sertoli cells do not contribute to the genetic makeup of the sperm, any effect of a mutation in them on a sperm cell would have to be epigenetic. Again this seems far-fetched. The sum of these considerations is that the increased cancer risk reported in the progeny of male workers at Sellafield is likely a statistical fluke or the result of a rare infection introduced when large numbers of people immigrated to the Sellafield area to build and operate the nuclear power facility.

A staple of horror movies is that radiation exposure will cause an irreversible genetic change in the human species. However, based on the data we discussed, there is no credible chance this will happen. At the risk of disappointing many horror movie fans, we must point out that these films foster unfounded fears of radiation. Here is why.

From 1910 to 1927, Thomas Hunt Morgan (1886–1945) and his colleagues Calvin Bridges (1889–1938), A. H. Sturtevant (1891–1970), and H. J. Muller (1890–1967) did pioneering work in genetics in their study of the common fruit fly *Drosophila* at Columbia University. They exposed flies to X-rays to see whether radiation would cause a mutation. Although they found many instances of single genetic changes, those changes were never permanent. A fly with no eyes might reproduce through several generations, but the progeny would revert to the normal two-eyed fly. Flies that were changed dramatically often died prematurely or could not reproduce. Most interesting, no mutations were beneficial; the mutations they observed over thousands of generations either had no discernible effect on the fly or were deleterious. Creationists have used these data to discredit the theory of evolution, arguing the data prove intelligent design. A more likely explanation is that flies had developed to an *optimal* state over billions of generations, such that additional mutations were statistically unlikely to be of benefit. A fly with a mutation making it resistant to a newly developed pesticide would, of course, be an exception, but this was not tested. These observations may also apply to humans. It is estimated that DNA mutations occur in 1 in every 100,000 genes. Because humans have 23,000 genes, we might expect one in every four sperm and eggs to contain a mutated gene. However, the frequency of medically important genetic mutations in human newborns is quite low or substantially less, and not all of these mutations are directly in DNA. Perhaps humans, like fruit flies, are optimally evolved. (Let's hope not.)

Some scientists think of cancer as a process of speciation, in which a cancer cell and its progeny are trying to become a new species within a person. This process resembles a distorted form of evolution on Earth. While humans have thousands of genes, a mutation in just one can cause cancer (although most cancer cells have many mutations). When you examine the chromosomes of cancer cells under the microscope, you often see an abnormal structure or number of chromosomes. This is odd because with 23,000 genes, a mutation in one or two or even ten would be so small as to be invisible. (They can be detected by more sensitive techniques like sequencing a person's DNA, a process that is now technically possible.) If a cancer is to develop from a DNA mutation, progeny of this cell must inherit the mutation of the parent cell. This is most likely to happen if the mutation that starts or causes a cancer alters the structure of a chromosome, the number of chromosomes, or both.

Humans share many, if not most, of our genes with other creatures. For example, we share about 70 percent of our genes with fruit flies and 98 to 99 percent with chimpanzees. What sets us apart from other species is the number of chromosomes we have compared to these other species, not the number of genes. (Some primitive creatures have more genes and chromosomes than us, such as zebra fish.) An analogy can be made to atomic number. The number of protons in an atomic nucleus determines what the element is, whereas the number of neutrons can vary, giving us different atomic weights of different isotopes, all of the same element. So 22 pairs of chromosomes plus either 2 X chromosomes (a female) or an X and a Y chromosome (a male), 46 in all, is what defines a human. Chimpanzees, in contrast, have 48 chromosomes, which is what makes them chimpanzees and not humans. There are, of course, exceptions to these rules, such as children with Down syndrome, who have an extra chromosome 21 (and therefore 47 chromosomes) or girls with Turner syndrome who have lost one of their X chromosomes (and so have 45 chromosomes). Using our analogy, we can consider these children human isotopes. Still the rule remains: humans have 46 chromosomes, and chimpanzees have 48. Because we share 98–99 percent of our genes with chimpanzees, the tremendous diversity we see in humans must come from fewer than 1 percent of our genes, probably fewer than 0.1 percent. This is amazing. Epigenetic features may also explain some of this diversity.

When the structure or number of chromosomes in a cancer cell is altered, we can think of the cell in some respects as a new species. This happens more than half the time in cancers in humans and thus is unlikely to occur by chance.

So the question is: Why is a change in chromosome structure and/or number so common in cancer? Where else is there a change in structure and number of chromosomes? The answer is in the distinction between species. So one might reasonably conclude that a cancer is a different species that develops within the affected persons and that its biological implication is to recapitulate the evolution of species. There was a species, and suddenly there is a new species—a reptile can suddenly fly (birds are derived from reptiles). This could result from a mutation in only one or a few genes, but the process is likely far more complex.

It is important to distinguish the evolution of a species and natural selection within a species. Consider pygmies and the very tall people of the Nilotic tribes of Sudan. Whether someone is taller or shorter is the result of natural selection. It has probably developed gradually, because being shorter or taller may be advantageous in different environments. If you are short, you can run faster under the low-hanging tree canopy of the forest and escape predators; if you are tall, your longer legs will allow you to run faster in the open. However, both tall and short people are *Homo sapiens,* and both have 46 chromosomes. This is variation within a species, not evolution of a species.

Back to radiation exposure, which can cause structural or numerical changes in a cell's chromosomes. Scientists look for these changes to determine the dose that a radiation accident victim has been exposed to. This is referred to as biological *dosimetry.* These structural (called *rearrangements* in genetics) and numerical chromosome changes occur because ionizing radiations tend to break both strands of the DNA double-stranded helix. These double-strand breaks can result in a *translocation,* whereby parts of two chromosomes, both of which have been broken by radiation, are incorrectly joined. Another example is a dicentric chromosome, where two copies of the same chromosome are joined after losing a small part of each. These incorrect recombinations can alter the number of chromosomes. For example, if two chromosomes join in a dicentric chromosome and lose the two small pieces, the cell will have 45, rather than 46, chromosomes.

We all start from one cell formed by the fusion of a sperm with an ovum, and all our organs and cells derive from this one cell. This cell is termed a *toti-potent stem cell* because it can give rise to every tissue, organ, and cell in our body. ("Toti" indicates universal potential.) Whether toti-potent stem cells persist in adult humans is controversial. Later in development some toti-potent stem cells mature into *pluri-potent stem cells* ("pluri" indicates many but not all potentials). These pluri-potent stem cells are restricted as to what organs, tissues, and cells in the body they can develop into. Further along in development are *multi-potent stem cells* ("multi" meaning many potentials), which have even less ability to turn into other cells. The process ends in *committed stem cells* that will irreversibly develop into a specific cell type, say a heart or liver cell. Our focus here is on pluri-, and multi-potent, and committed stem cells found in our bone marrow. These cells produce the mature blood cells we need in order to live, including red and white blood cells and megakaryocytes (which mature into platelets, which help stop bleeding).

It is possible to give a mouse a dose of radiation that will kill most all of its bone marrow stem cells. The mouse will die of infection and bleeding within a few days or weeks because the bone marrow cannot produce the mature blood cells it needs to survive. Fortunately, it is possible to rescue the radiated mouse by transplanting into it bone marrow stem cells from another mouse. The recipient mouse can recover completely, living on the bone marrow of the donor mouse. If the radiated mouse was male and the donor mouse female, the bone marrow and blood cells of the radiated mouse will now be female, whereas the skin cells will remain male. Interestingly, using genetic marking by a RNA virus (a retrovirus such as the AIDS virus), it is possible to show that bone marrow recovery of the radiated mouse comes from only one or a few cells. These cells are the multi-potent stem cells.

Because our blood cells have a short lifespan—120 days for red blood cells, 10 days for platelets (not really cells but parts of cells), and a few hours or days for granulocytes—our bone marrow needs to produce more than 3 billion cells every day throughout our lifetime. To do this, our bone marrow relies on an amplification system starting with stem cells. And our stem cells must last us a lifetime, because if we run out of them, we die.

We live in a radioactive environment. Consequently, our bone marrow stem cells are at constant risk of acquiring mutations from radiation. So one necessary step in the evolution of life was to protect bone marrow stem cells from background radiations. Cosmic and terrestrial radiation is the greatest threat to life from a mutation perspective. In the early aeons of Earth, such radiation was much stronger than it is today, and the earliest organisms had to withstand these radiation forces.

Life probably started in water, which is a good shield against cosmic and terrestrial radiations. (Water is used in nuclear power facilities to protect workers from radiation from spent fuel rods.) This shielding effect may have influenced the evolution of life on Earth. Consider frogs. A frog living under water is shielded from cosmic and terrestrial radiation, so it does not matter where its blood-producing stem cells are, because its body does not need to protect them from radiation. The frog's blood-producing cells may be in its liver, spleen, or kidney, because, again, they are shielded from radiation by being under water. As frogs evolved and spent more time on land, there was no water to shield the blood-forming stem cells from cosmic and terrestrial radiation, placing these blood-forming stem cells at risk for mutations. When frogs became entirely terrestrial, the situation was even worse, because the blood-forming stem cells were no longer protected by any water and the risk of mutations was even greater.

A good way to protect blood-forming stem cells from cosmic and terrestrial radiations is to put them inside a bone. Bone is made up of a form of calcium called hydroxylapatite, a good shield from radiation. This ability to block radiations is why bones appear white (opaque) on an X-ray; X-rays have a hard time passing through them. Most human bones are hollow, and our bone marrow cells, and hence our blood-forming stem cells, live in the bone marrow cavity of our pelvis, ribs, and vertebrae. From a structural viewpoint, it might have been better to have solid rather than hollow bones because they provide greater strength, but from a radiation-protection point of view, it is better to have bone with a hollow core. Fish, which live their whole life under water, have solid bones. Their blood-forming stem cells are in their kidneys. These data suggest that exposure to cosmic and terrestrial radiations may have played an important role in how humans and other creatures evolved. We may wonder why whales and dolphins living in water have their bone marrow stem cells in bone cavities. This is because they derive from terrestrial ancestors.

But the data go only so far. An argument can be made that reproductive cells are at least as important as blood-forming cells, and probably more so. Thus if blood-forming stem cells are in bone to protect them from radiation, wouldn't one expect human testes and ovaries to also be protected in bone? To our knowledge this is not so. (Please treat the notion of hollow bones and bone marrow as a hypothesis rather than a fact.)

IS IRRADIATED FOOD DANGEROUS?

It is difficult to arrive at a worldwide figure for incidence of food poisoning, but the World Health Organization estimates that more than 2 billion cases occur each year, and an estimated 2 million people die from infections of the gastrointestinal tract such as typhoid fever and cholera, many of which can be traced to bacteria in food and water. According to the U.S. Centers for Disease Control and Prevention (CDC), in the United States bacteria in food cause an estimated 48 million cases of food poisoning annually. Around 130,000 people are hospitalized for food-borne illnesses, and nearly 3,000 die from eating food contaminated by bacteria such as *E. coli*, *Salmonella*, and *Listeria*.

Yet many people say they would rather be exposed to these bacteria than eat food that has been radiated to kill them. This fear mostly stems from the misconception that radiated food can become radioactive. The source of the gamma or X-rays used to radiate the food never comes in direct contact with the food, so the food is not made radioactive by these rays; nor does it contain extra radioactivity. (All food is normally radioactive.) It is as physically impossible for food treated with gamma rays to become more radioactive as it is for your teeth to become more radioactive after a dental X-ray. (You could make food radioactive by exposing it to neutron radiation and thereby activate sodium-24, but neutrons are never used for food radiation.)

The CDC says that 90 percent of all illnesses resulting from food-borne pathogens are caused by just 7 of the 31 that are known: *Salmonella, Norovirus, Campylobacter, Toxoplasma, E. coli* O157:H7, *Listeria,* and *Clostridium.* Just as pasteurizing dairy products and pressure-cooking canned foods kill germs (when done properly), exposing food to ionizing radiation kills bacteria and parasites that otherwise could make you ill. (Commercially canned foods are not guaranteed safe, as shown in the 1971 death from botulism of a man who ate tainted vichyssoise.) Whether it comes from gamma rays released by cobalt-60 or cesium-137, electron beams, or X-rays, the high energy of ionizing radiation breaks the chemical bonds that bacteria and other microbes need for growth, leaving them unable to multiply and cause food to spoil or make people ill. It allows food to stay fresh longer. Irradiated food is a staple for astronauts in space. Many brands of bottled water we drink are sterilized by UV radiations, although the label may not always say so.

Every method used to process food—including storing it at room temperature for a few hours after it is harvested—can lead to a loss of some vitamins. Sterilization using radiation, in contrast, results in loss, if any, at a level so low as to be undetectable or irrelevant. So the high dose of radiation used to increase shelf life lessens food's nutritional value no more than cooking or freezing it does.

Groups opposed to food irradiation point out that gamma rays alter the molecular structure of the food, forming charged particles known as *free radicals* that interact with molecules in the food to then form "radio-lytic products" like mutations in DNA. A particular concern is that irradiation causes the increase of benzene levels in meat. Certainly in a large enough quantity benzene is a carcinogen, but it is already present in small amounts in many foods, water, and air. In fact, there is more benzene in normal milk than in irradiated meat. And you also get a substantial extra dose of benzene when you fill your car with gasoline (which contains small amounts of benzene) or when you stand on a busy street corner in New York or Cairo during a traffic jam.

To keep perspective, it is important to recall that all food treatment causes chemical changes. Freezing food causes what could be called "cryolytic products." Cooking causes more changes than irradiation—the taste and smell of cooked food are produced from chemical changes brought about by heat. Barbecuing a steak produces many known carcinogens on the steak surface from the interaction with charcoal, a hydrocarbon. That increases global warming.

The American Medical Association's policy toward food irradiation is that it is "a safe and effective process that increases the safety of food when applied according to governing regulations." The U.S. Food and Drug Administration (FDA), the World Health Organization, the International Atomic Energy Agency, and scientific organizations in many countries confirm the safety of food irradiation. Irradiated food is permissible in more than 50 countries, with a variety of clearances: the European Union, for instance, allows only irradiated spices, whereas Brazil permits all foods to be irradiated. The FDA has approved irradiation to, among many other uses, control sprouting of food grown in the ground (such as onions, carrots, potatoes, and garlic); delay the ripening of bananas, mangos, papayas, and other noncitrus fruits; and kill insects in wheat, potatoes, flour, spices, tea, fruits, and vegetables. The FDA also approves irradiation of pork to control trichinosis and to eradicate *Salmonella* and other harmful bacteria in chicken, turkey, and other fresh and frozen uncooked poultry.

Food irradiation has the potential to save millions more lives than it harms, especially since it very probably does *no* harm. It is especially important in developing countries where most cases of food poisoning occur.

CHAPTER 6

RADIATION AND MEDICINE

MAMMOGRAMS

Breast cancer is the most common cancer in women worldwide. The World Health Organization reports that it caused about 458,000 deaths in 2008, about 14 percent of all cancer deaths in women. There are around 180,000 new cases of breast cancer in the United States annually and about 40,000 deaths. Breast cancer is the second leading cause of cancer death in American women, after lung cancer. (Breast cancer was the leading cause in women until 2000, when the effects of women smoking cigarettes, which became popular in the 1920s and increased in the following decades, made their fatalities similar to men's.) The American Cancer Society estimates that an average woman's chances of developing breast cancer in a normal life span is 10 to 15 percent, depending on nationality, family history, age of first pregnancy, breast-feeding, alcohol use, and other factors. Some women—about 5 percent—with a heritable genetic risk for breast cancer, such those who have BRCA1 and BRCA2, have a more than 80 percent likelihood of developing it.

(Men can develop breast cancer, but the disease is about 100 times less common.)

Survival of women with breast cancer correlates with the extent to which the cancer has spread and other factors, such as whether the cancer cells respond to estrogen and progesterone, and whether they have certain specific genetic abnormalities. Women whose breast cancer is small and remains within the breast (and possibly the local lymph nodes) are much more likely to be cured than women who have large cancers in their breast or whose cancer has spread beyond the breast or to many lymph nodes. The main reason for this correlation is mostly simple: breast cancers that remain in the breast can be cured by surgery (with or without other therapies like radiation), whereas metastatic cancer is mostly incurable. But another reason for this correlation is biological. Some breast cancers spread outside the breast soon after they develop, often before they can be detected, whereas other cancers remain in the breast for a very long time despite growing larger. The latter have a much better prognosis than the former. There are more than 2.5 million breast cancer survivors in the United States.

Regardless of these complex considerations, the correlation between early detection of breast cancer and increased likelihood of cure with less intensive therapy has resulted in considerable efforts in early breast cancer detection or screening. And because breast cancer is so common, this involves screening millions of women.

Mammography is the only screening test for breast cancer that has been shown to reduce deaths from breast cancer. (The effectiveness of breast self-examination is controversial.) Each year about 40 million American women have a mammogram in the hope of early detection of breast cancer. Their greater hope, of course, is that no abnormality will be found.

Mammograms use low-energy X-rays to detect abnormalities in breast tissue. The whole body dose from the average mammogram is about 0.2–0.4 mSv, which is about one-tenth of a woman's average annual dose from background radiation. The radiation dose from a mammogram is considered to be too low to pose substantial individual risk; however, as with any use of ionizing radiation, repeated exposures can, at least in theory, cause or contribute to the development of cancer, including breast cancer. And some women with genetic abnormalities that predispose to breast cancer and who are already at high risk of breast cancer (such as those with the BRCA-2 mutation) seem at special risk to develop more breast cancers if they receive chest X-rays or mammograms.

The decision to recommend any medical screening procedure, including mammograms, is based on a delicate balance of estimating potential benefit and risk. So there is heated controversy regarding the age women should begin having annual or biannual mammograms for early detection of breast cancer, and at what age they should stop.

As we discussed, exposure to high doses of ionizing radiations causes breast cancer, leukemia, and other cancers. In the A-bomb survivors, breast cancer was one of the cancers most likely to be caused by ionizing radiation. The likelihood that radiation will cause or contribute to breast cancer development is greatest in women who are young at time of exposure, and this risk decreases with increasing age. Whether a very low dose of radiation, like that associated with a mammogram, can cause breast cancer and leukemia is controversial. Most scientific and regulatory agencies and scientists agree that to protect the public, we should assume that even the smallest radiation dose can potentially cause cancer. Others disagree, some strongly.

So we are left with a complex decision as to what recommendation to make. The earlier the age women begin having mammograms, the greater the possibility that some may develop breast cancer or another cancer or leukemia from radiation exposure. But paradoxically, women who are at the highest risk of developing breast cancer—perhaps because of an inherited genetic susceptibility or strong family history—and are therefore most likely to benefit from screening mammograms, are those most likely to be at risk of developing breast cancer from low-dose radiation.

The bottom line of this complex calculus is that breast cancer screening using mammograms saves lives and reduces the amount of therapy that some women need because their cancers are detected early when they remain in the breast; they may receive only surgery and local radiation therapy rather than anticancer chemotherapy or hormone therapy. Between 15 and 20 percent of breast cancer deaths, perhaps more, are prevented by screening mammograms. So this issue is not yes or no; it is when, and how often.

Here controversy prevails. The U.S. Preventive Services Task Force (USPSTF) and the CDC recommend that women with no risk factors for breast cancer have a screening mammogram every two years between the ages of 50 and 74. In contrast, the National Cancer Institute, American Cancer Society, and several other professional organizations recommend annual mammograms beginning at age 40 and continuing as long as the woman is in good health. In a USPSTF analysis, beginning screening at age 40 saved about 5 percent more lives than beginning screening at age 50, but it was associated with many more inaccurate and ultimately incorrect diagnoses. (Abnormalities on mammograms thought to be cancers were shown to have a cause other than cancer. This result is referred to as a false positive.) Although a correct diagnosis was finally made, these false-positive mammograms incurred considerable physical, psychological, and economic costs. Some women required one or more breast biopsies; others developed a perhaps irrational, sometimes disabling, fear of developing breast cancer. Also, screening annually or every two years had similar reductions in breast cancer deaths but fewer false-positive breast cancer diagnoses. Recently, several states enacted laws requiring physicians to advise women with dense breasts that mammography may not be sufficient to detect breast cancer and that additional tests, like ultrasound and MRI, may be needed. The problem is we lack good data on the added benefit of these procedures in most women with dense breasts. We regard this as an unfortunate intrusion (confusion) of politics and science.

Ionizing radiations play a complex role in breast cancer. They can cause it, they can be used to diagnose it early and save lives, and they can be used to treat it and save lives. From a radiobiology perspective, screening mammograms are a good example of the benefits (early diagnosis of breast cancer) exceeding the potential risks (exposure to ionizing radiations).

LUNG CANCER SCREENING

Lung cancer is the most common cause of cancer death in men and women worldwide: an estimated quarter-million cases will occur in the United States in 2012, resulting in about 165,000 deaths. Most lung cancers are diagnosed at an advanced stage, when they are no longer curable or even effectively treatable. Because of the high death rate (the average survival is less than one year from diagnosis), there is considerable interest in early diagnosis. The situation here is rather different from that of breast cancer. Although lung cancer is common, not everyone is at risk; the lifetime risk of lung cancer in American males who are nonsmokers is less than 1 percent, and it is even lower in females. But males who are heavy smokers have a more than 25 percent lifetime risk, so lung cancer screening is best directed at high-risk smokers (or former smokers) rather than at all people of one sex, as is the case in breast cancer.

Prior methods of lung cancer screening used conventional chest X-rays and analysis of sputum samples for abnormal cells. Results were disappointing, and there are no convincing data that these interventions save lives. Recently, a special radiological study—a low-dose helical CT scan of the chest—was tested as a lung cancer–screening technique in people at high risk: smokers (or former smokers) with a history of smoking for more than 30 pack-years, such as two packs a day for 15 years or one pack per day for 30 years. A study of more than 50,000 persons showed that the group screened with low-dose helical CT scans had a 20 percent decrease in lung cancer deaths over those who were screened with conventional chest X-rays.

But what does this really mean? One would have to do helical CT scans biannually over six years in about 30,000 high-risk people for three years to prevent about 60 lung cancer deaths. Nevertheless, based on these data, the American Society for Clinical Oncology, the American Lung Association, and the American College of Chest Physicians now recommend screening for persons at high risk of lung cancer. Other health organizations such as the USPSTF have yet to endorse this recommendation. We should recall about half or more of lung cancers occur in nonsmokers or persons smoking less than 30 pack-years. These people were not included in the above study because, although they contribute a large proportion of lung cancer cases, the risk in any one person is so small that the potential benefit-to-risk ratio of screening is unfavorable. Even with the smoker population, risks of developing lung cancer vary greatly. Consequently, the likelihood screening will prevent a lung cancer death in a person varies more than tenfold. Therefore, recommendations regarding lung cancer screening should be individualized.

The impact of a successful lung cancer screening program is anticipated to be small—less than 5 percent of lung cancer deaths could be prevented. In contrast, the benefit-risk of the radiation involved is favorable, as only persons at high risk of developing lung cancer are to be screened. People selected for screening have a much higher risk of lung cancer than the average woman has of developing breast cancer. The conclusion that people at high risk for lung cancer should have screening radiological studies remains controversial but presently favors screening.

COMPUTED TOMOGRAPHY (CT) SCANS

Allan Cormack (1924–1998), a South African–born physicist, was so entranced by the stars as a boy that he studied physics in order to become an astronomer. In time, however, the appeal of the vastness of space was replaced by a fascination with the subatomic universe of particle physics and a parallel interest in X-ray technology. These interests, combined with the serendipity that is so much a part of science, led him to figure out how to make a three-dimensional image of something inside a body. He was amazed that no one had thought to do it sooner.

In 1955, fresh from two years of postgraduate study at Cambridge University, Cormack was asked by a Cape Town hospital to fill in part time in the radiology department, a post for which there was no competition, as he was the only person in the city with training both in physics and in handling radioisotopes. His task was to figure out how to deliver an exact dose of X-rays to a specific point in the body. That required determining what amounts of X-ray energy are absorbed by particular parts of the human body. Only after he began the work did he appreciate how crude X-ray images really are; they show everything in their path layered on one another in two dimensions.

This was sixty years after Röntgen's first X-ray. Cormack assumed that Röntgen had deciphered how much energy is absorbed by tissue of various density, so the task would be easy. All he had to do was make a map of the body by taking X-ray images from many angles, then use a formula of triangulation to create a high-definition picture. But to his surprise, he could find no paper that showed that either Röntgen or any subsequent scientist had done the math, nor apparently even tried. Over the next several years, Cormack refined the mathematics that would enable radiologists to make one clear image from many X-rays taken at angles and directions that crisscross the body. (Years later he would find that work similar to his own had been done in World War I by the Austrian mathematician Johann Radon, who later lived in England.) Cormack polished and tested his theories using unsophisticated methods; then, happy to leave the rest to engineers to perfect, he published two papers in the *Journal of Applied Physics,* in 1963 and 1964, and went on to other work.

While taking a stroll in the countryside in 1967, Godfrey Hounsfield (1919–2004), a British electrical engineer, wondered if it would be possible "to determine what was in a box by taking readings at all angles through it." An expert on radar who had also done groundbreaking work in computer memory design, Hounsfield quickly decided that using a computer to put together many X-ray images would show not only what was in a box but what was in a human skull as well. As oblivious of Cormack's work as Cormack was of Radon's, Hounsfield constructed the first CT machine. In 1971 an improved version was used successfully in a hospital and became known as body section röntgenography or computed axial tomography. ("Tomography" is derived from the Greek *tomos,* "slice," and *graphein,* "to write.") CT was soon hailed as the greatest advance in radiology since Röntgen's initial work. The Hounsfield scale is the standard measurement for evaluating CT scans, and he and Cormack shared the 1979 Nobel Prize in Physiology or Medicine.

CT scans have become a staple of medicine; there are about 6,000 CT machines in the United States. The American Cancer Society estimates that in 2007 Americans underwent an estimated 72 million scans (in 1980 there were 3 million), and the annual number is growing. This is troubling, because although a CT scan is an alluring and useful diagnostic tool, during a single rotation of the scanner around the body, a computer takes 175 to 200 (or more) two-dimensional X-ray images of the body interior and turns them into a high-contrast, three-dimensional image. That large a dose of radiation, especially at one time, can be dangerous, because of the risk of causing cancer.

A study published in *The Lancet* in August 2012 of children having CT scans for diverse reasons reported they had an increased risk of leukemias and brain cancers. The benefit of a CT scan compared to a conventional X-ray is enormous; it enables physicians to see abnormalities missed by X-ray and to pinpoint their location. But CT scans also deliver much more radiation than a conventional X-ray. Consequently, CT scans should be used judiciously. Health experts consider about one-third of CT scans unnecessary, but they are done because physicians do not always calculate whether the benefit-to-risk ratio of radiation exposure is favorable. Some scans are done either because the patient demands it or because the physician wants to reduce risk of litigation, or both.

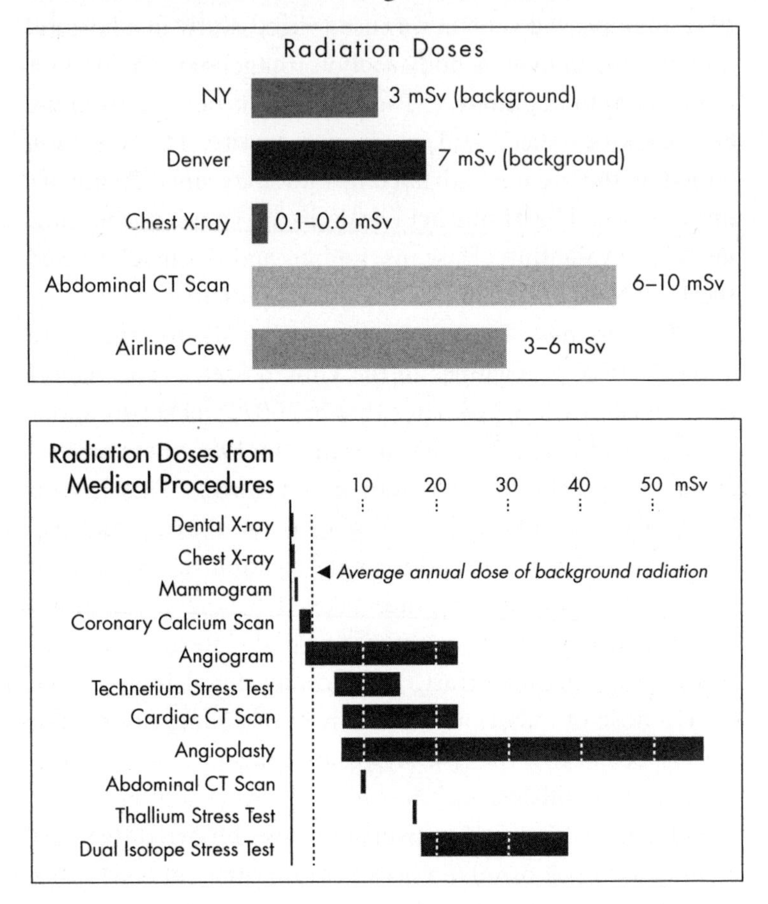

POSITRON EMISSION TOMOGRAPHY (PET) SCANS

In a PET scan, fludeoxyglucose (FDG), a radioactive form of glucose containing fluorine-18, is injected into the blood. PET scans look for metabolic abnormalities in the body where glucose is being concentrated and burned. Like normal glucose, FDG is concentrated in metabolically active sites like infection or cancer. The positron that FDG emits can be detected by a sensor. Although a CT scan or magnetic resonance imaging (MRI) shows the structure of organs but not their function, a PET scan gives a two-dimensional sense of where the increased metabolic activity is. A simultaneous CT scan or MRI gives a three-dimensional pinpoint of the cancer or infection. The person must wait about an hour after injection for the FDG to disperse throughout the body and concentrate at the site of glucose metabolism. Fluorine-18 has a half-life of about 2 hours. The positive positron that it emits travels less than 1 four-hundredth of an inch. In this short span it loses energy to the point where it can interact with an electron. PET scans are also called *annihilation therapy* because that interaction annihilates both the electron and the positron and produces a pair of gamma photons that can be detected with an appropriate radiation detection device. As the tracer decays, the scan makes a record of tissue concentration. Most of the fluorine-18 is flushed from the body over 6 to 24 hours.

A PET scan is noninvasive, but exposes a person to ionizing radiations. The dose of radiation is substantial, usually around 5–7 mSv, twice one's usual annual background radiation dose. However, in modern practice, a combined PET/CT scan is almost always performed, and for PET/CT scanning, the radiation exposure is considerable—20–25 mSv. This is 7 or 8 times the normal annual background radiation dose. There is both internal and external radiation in a combined PET/CT scan—internal from the radioactive tracer, and external from the X-rays passing through the body from the CT scan. It is essential that physicians calculate the potential benefit-to-risk ratio of performing a PET or CT scan before requesting such a study. People should ask their physician why the test is being done, precisely how they will benefit, and how much radiation they will receive. They can then make an informed decision about whether to agree to the study. If your physician or radiologist cannot answer your questions, you are probably in the wrong place.

RADIATION THERAPY

Radiation therapy is the use of relatively high doses of high-energy ionizing radiations to treat diseases, especially cancers. In chapters 3 and 4, we discussed several examples of radiation therapy, including use of high doses of iodine-131 to treat thyroid cancer, yttrium-90, linked to an antibody to treat lymphomas, and strontium-90 to treat pterygium of the eye. Many other forms of radiation therapy are used to treat cancers. Some cancers are best treated and even cured using radiation therapy.

Radiation therapy contributes disproportionately to the average man-made radiation dose received by the public. Although relatively few people in a population receive radiation therapy, the dose is extremely high. From a radiation biology perspective, the key variable for predicting the health consequences of radiation exposures is *cumulative effective dose.* You are already familiar with individual effective dose expressed in mSv. Imagine that 1,000 people each receive an effective dose of 1 mSv of ionizing radiation. The cumulative effective dose is 1,000 people times 1 mSv, or 1,000 person · mSv. The cumulative effective dose is the same if instead one person receives 1,000 mSv and the other 999 receive no radiation. Now consider radiation therapy, where one person may receive 60,000 mSv. This is the same cumulative effective dose as 60,000 people receiving 1 mSv. So you can see why each person receiving high-dose radiation therapy contributes disproportionately to calculations of the average radiation dose to the population and thereby to the estimate of health consequences of excess radiation to a population.

Although ionizing radiations can cause cancer, very high radiation doses irreversibly damage DNA and kill cells—the exact opposite of cancer. So high-dose radiation can be used to treat cancers. Typically, the types of radiation used include X-rays, gamma rays, and charged particles like protons (some calculations ignore radiation therapy in determining average population dose as it applies only to people with cancer), neutrons, and heavier ions such as carbon and neon. Sometimes alpha particles are used. Drugs called radio-sensitizers seem to increase the ability of radiation to kill cancer cells.

Radiation therapy can be external, when the radiation source is a machine outside the body, or it can be internal (also called *brachytherapy*), as when radioactive pellets or seeds of cesium-137 are placed inside the body near the cancer to treat people with prostate and breast cancers. Finally, radionuclides can be injected into the blood, such as high doses of iodine-131 to treat thyroid cancer, or by attaching a radionuclide to an antibody reacting with lymphoma cells: the antibody pulls the radionuclide to the lymphoma cells, so the radiation dose to the rest of the body is small while the dose to the lymphoma is high. It's like a nuclear weapon, where the antibody is the targeting missile and the radionuclide is the warhead. Injecting radionuclides into the body, usually intravenously, is referred to as *systemic radiation therapy.*

In the United States, about half of all people with cancer other than skin cancers—that's more than half a million people each year—receive some form of radiation therapy. Some receive only radiation therapy, whereas others receive radiation therapy combined with surgery and/or anticancer drugs or hormones.

The damaging effects of ionizing radiations are not specific to cancer cells. Normal cells near the cancer cells are also killed by radiation scatter, and there are bystander effects whereby substances released by cells affected by radiation can cause biochemical changes in nearby non-irradiated cells. Most types of cancer cells are thought to be more sensitive to radiation-induced killing than normal cells, perhaps because normal cells are better able to repair radiation-induced damage. To take advantage of this difference, radiation therapy is usually given in fractions; the total dose is divided into smaller doses and given over several days or weeks. The notion is normal cells will be able to repair radiation-induced damage during the interval between doses, while the cancer cells cannot. Fractionation of radiation allows a higher cumulative dose of radiation to be given than when the total dose is given in one fraction. Some cells are simply more sensitive to radiation, as with xeroderma pigmentosum (see chapter 4), where there is a defect in repair of UV radiation–induced double-stranded DNA breaks. This is what gives XP children a high risk of developing skin cancers. One can imagine that an increased sensitivity to radiation might characterize some cancers. Radio-sensitizers seem to increase the efficiency with which radiation kills cancer cells.

Radiation therapy can be used to eradicate a cancer or to prevent a recurrence. For example, women who have a partial breast resection (lumpectomy) for breast cancer typically receive radiation therapy afterward to kill potential residual cancer cells missed by the surgery. This approach is effective in reducing risk of recurrence. Both external and internal radiation therapies are used, depending on the cancer location, size, and other data correlated with an increased risk of the cancer recurring locally.

Radiation therapy is sometimes used to reduce the size of a cancer, without intending to cure the person. Such palliative therapy is especially important if, for example, a large lung cancer blocks an airway and interferes with breathing or erodes into an important blood vessel and causes death from bleeding, or if a cancer of the larynx or esophagus is preventing breathing or eating. This type of palliative radiation therapy is often used to treat cancers that spread to the brain because many anticancer drugs do not cross into it. Palliative radiation therapy is also used to treat specific sites in the bones of people with multiple myeloma and other cancers who have collapsed vertebrae or other bones from invasion of cancer cells. Sometimes these bone metastases compress the spinal cord and can cause paralysis unless quickly treated with radiation therapy.

Successful radiation therapy requires precise definition of the site(s) to be included in the radiation port or field. This is usually accomplished by a process termed *simulation.* CT and/or PET scans and MRI are used to define the boundary between the cancer and normal cells. Next the radiation oncologist determines the doses to be given to the cancer and to the normal surrounding tissues and the most effective path of the radiation beams to achieve this. Because of this precise targeting, the persons receiving radiation therapy must be in exactly the same position for each dose fraction. This is accomplished by several techniques, such as a mask on the head or tattoo marks on the skin or imaging immediately prior to or during the radiation (*image-guided radiation*).

Different tissues and organs have different sensitivities to radiation. As we mentioned, the bone marrow, the gastrointestinal tract, and the skin are extremely sensitive to radiation because cells in these tissues are rapidly dividing. Other organs and tissues, such as the brain and kidney, can tolerate much higher radiation doses without irreversible damage. The radiation oncologist needs to consider the radiation sensitivity of the cancer and that of the surrounding normal tissues and organs to determine the best dose for each type of cancer and for each person. This is because some cancers are more sensitive to killing by radiation therapy than others. Also, the location of a type of cancer may differ from person to person, so the surrounding normal organs and tissues that need to be protected from radiation damage will differ.

Ideally, the treatment will deliver a high radiation dose to the cancer as defined in the simulation. It is also desirable to irradiate a small volume of normal tissue immediately next to the cancer because some cancer cells not detected by the simulation studies may have spread to these sites. (Cancer cells that may have spread but are undetectable by conventional techniques are called *micro-metastases.*) Moreover the person receiving radiation therapy may not be in precisely the same position at every therapy session. The dangerous possibility is diminished with newer technology that employs image guidance treatment.

Different types of particles and waves are used in radiation therapy settings. External-beam radiation therapy typically uses photons (see Introduction). Often these photons are generated in a linear accelerator called a LINAC that uses electricity to create a photon beam. There are several forms of external beam radiation. The most common, *three-dimensional conformal radiation therapy,* uses sophisticated computer programs to precisely shape the radiation beam to the desired target area. *Intensity-modulated radiation therapy* uses small devices to shape the radiation beam, allowing changes in radiation intensity during treatment to avoid normal tissue. And *image-guided radiation therapy* uses imaging studies like CT scans or MRIs, repeated throughout the treatment, to allow the radiation field to be adjusted to again maximize the dose to the tumor and minimize normal tissue damage. All of these types of radiation therapy have the goal of increasing specificity: applying more radiation to the cancer and less to normal tissues. Furthermore, all types of radiation delivery can take advantage of image-guided radiation. These techniques have already shown improvements in survival rates for head and nect cancer and prostate cancer, with dimished late toxicity in children and patients with brain tumors.

Sometimes physicians want to deliver a very high radiation dose to a small, precisely defined area, such as the pituitary gland in the brain. They may apply a technique called *LINAC-based stereotactic radiosurgery* or a gamma knife. It uses gamma rays to pinpoint a target area and is meant to avoid surgery deep in the brain. When this technique is used to target cancers outside the brain, it is referred to as *stereotactic body radiation therapy,* another LINAC-based treatment.

Some forms of external-beam radiation therapy use charged particles like protons instead of photons. The theoretic advantage is that these particles deliver their energy directly into the target whereas photons deposit some of their energy en route to their target. This means protons are able to avoid damage to normal tissues more effectively than photons. Photon generating machines are 20 to 30 times more expensive than LINACs and thus access will be limited. The newer LINAC technology avoids the need to use protons for most adult tumors. Another form of radiation therapy uses beams of electrons. Since electrons cannot penetrate the body effectively, most electron beam therapy is directed toward skin cancers or superficial tissue. An even newer form of external beam charged particle therapy uses ions of carbon, lithium, boron, and other elements.

These forms of external radiation therapy should be contrasted with radiation therapy in which radionuclides are placed inside the body at the site of a cancer. These therapies have different names, depending on whether the radiation source is placed within the cancer, within a body cavity, or within a specific site like the eye. Brachytherapy uses radionuclides sealed in small capsules called seeds that are sited using needles and catheters. As the radionuclides in the seeds decay, they release radiation to the surrounding tissues. This allows a very high radiation dose to be precisely delivered while minimizing damage to normal tissues. Sometimes the seeds are removed after therapy is completed, but sometimes they are left in place forever.

In some forms of systemic radiation therapy, the cancer patient swallows a radionuclide, or it is injected. Here, the radionuclide needs to be targeted to the cancer. For thyroid cancer, the injected iodine-125 or -131 is taken up selectively by the thyroid gland, minimizing radiation damage to other tissues. Another technique is to attach the radionuclide to an antibody specific for the cancer cells. A third type of systemic radiation therapy uses radionuclides such as samarium-153 and strontium-90 that concentrate in bone to treat bone metastases.

The schedule and timing of radiation therapy varies considerably. Some people receive it before surgery, with the objective of shrinking the cancer, to make the surgery less extensive and possibly more effective. This is sometimes referred to as *neoadjuvant radiation therapy.* Others receive radiation therapy after surgery to decrease the likelihood of the cancer's returning locally. This strategy is usually used after a lumpectomy and is sometimes referred to as *adjuvant radiation therapy.* Radiation therapy given during surgery—*intraoperative radiation therapy*—can use external or internal radiation sources. Radiation therapy given in conjunction with anticancer drugs is called *chemoradiotherapy.*

It is commonly used to treat several cancers, including Hodgkin disease and other lymphomas.

People receiving radiation therapy are often concerned that they will become radioactive. This never happens with external radiation therapy. By contrast, internal radiation therapy can make a person radioactive for as long as the radionuclide remains in the body and continues to decay. Once the seeds are removed or decay is over, the person is no longer radioactive. Also, because the radioactive seeds are usually placed deep inside the body, very little radiation reaches the body's surface. In some instances however, internal radiation can release radioactive substances that can be detected in the urine, saliva, sweat, and other body fluids. These materials must be handled carefully.

Radiation therapy is an effective anticancer treatment, but it can also produce adverse effects, which we summarize next. Despite the adverse effects, the benefit-to-risk ratio is sufficiently favorable to make radiation therapy an important anticancer treatment.

LONG-TERM EFFECTS OF RADIATION TREATMENT

Radiation therapy has different long-term effects among people—some have none, others have serious or even life-threatening long-term effects. We generally assume that radiation therapy to one part of the body does not increase the risk of developing cancer elsewhere. But there is an increased risk of a new cancer at the site of prior radiation therapy, although such cases are rare. Long-term effects of radiation therapy include cataracts, cavities and gingivitis, heart problems, hypothyroidism, infertility, intestinal problems, lung disease, memory problems, and osteoporosis (bone loss), depending on the site irradiated. Radiation to the chest can cause later problems with heart and lungs; radiation to the abdomen can cause bladder, bowel, or sexual problems. New technology allows for radiation to be more precisely targeted than before, so fewer healthy cells are damaged. As the precision of radiation using fixed and rotational fields advances, physicians may be able to increase cures and decrease complications. It is important that the new technologies be evaluated with rigorous long-term studies and controlled clinical trials.

CHAPTER 7

BOMBS

NUCLEAR WEAPONS

Nuclear weapons are explosive devices whose force comes from a nuclear reaction: fission (atomic bombs) or fission and fusion (thermonuclear or hydrogen bombs). Fission and fusion reactions—as in the detonation of uranium-235 and/or plutonium-239—release huge amounts of energy compared to conventional weapons. The atomic bomb exploded over Hiroshima (Little Boy, perhaps named for Franklin D. Roosevelt or J. Robert Oppenheimer) is estimated to have released an amount of energy equivalent to about 16,000 tons (16 kilotons) of trinitrotoluene (TNT). Many modern nuclear weapons are 500 times more powerful, equivalent to 10,000 kilotons. One would have to explode 10 million tons of TNT to release the same amount of energy as in a single modern nuclear weapon.

Although Little Boy, a uranium-235 bomb, weighed almost 5 tons, the amount of uranium-235 it contained was only about 175 pounds (80 kilograms), and the amount of matter converted into energy was only 0.0015 pounds, about the weight of 30 grains of rice (about 700 milligrams). The Nagasaki atomic bomb, Fat Boy (perhaps named for Winston Churchill or Brigadier General Leslie Groves, who headed the Manhattan Project), contained plutonium-239 rather than uranium-235. Fat Boy's explosive force was equivalent in energy to about 21 kilotons of TNT. It instantly killed 60,000 to 80,000 of the 240,000 inhabitants of Nagasaki and injured 80,000, fewer than Little Boy. Fat Man was detonated over Nagasaki because the sky over Kokura City, the intended target 96 miles away, was clouded over. There were fewer fatalities because the bomb was not detonated at the optimal altitude and the hilly terrain of Nagasaki (it resembles San Francisco and Genoa) shielded residents in low-lying areas from the worst effects of the explosion.

Casualty numbers from the atomic bombs are startling, but it is important to compare their effects with those of conventional weapons. As we detailed, the numbers of people killed by the atomic bombs are not vastly different from the numbers of people killed by the firebombings of Tokyo (probably more than 100,000) and Dresden (about 25,000) during World War II using conventional bombs.

There is considerable misunderstanding about how nuclear weapons kill or injure people. They kill most people in the same way conventional bombs do: by generating concussive forces (shock waves) and by igniting superfires. As we saw in chapter 2, these effects accounted for about 90 percent of deaths in Hiroshima and Nagasaki. About 50 percent of the energy released by the A-bombs was blast energy, about 35 percent was thermal energy, and only about 15 percent was radiation, most of it neutrinos that did not contaminate the area. What makes nuclear weapons distinct from other weapons is the large number of immediate fatalities from a single source. Radiation is unique to nuclear weapons, and it is estimated to have caused about 10 percent of the early (but not immediate) deaths in Japan. Also, while radiation exposure from the atomic bombs unquestionably increased cancer risk in the survivors, the proportion of cancer deaths attributed to A-bomb-related radiation is only 8 percent of all cancer deaths over the past 65 years (about 550 out of 6,300 cancer deaths). This is because most of us have a 38 to 45 percent chance of dying from cancer without excess radiation exposure, depending on gender.

Since the atomic bombings in August 1945, more than 2,000 nuclear weapons have been detonated by the United States, the successor states of the Soviet Union, the United Kingdom, France, China, India, Pakistan, probably Israel, and possibly North Korea. These countries are estimated to have about 20,000 nuclear warheads, of which about 25 percent are immediately deployable. Most of these weapons are fission-type bombs, but some countries also have fusion-type weapons, commonly referred to as hydrogen (H-)bombs or thermonuclear weapons.

Fission-type nuclear weapons use conventional explosives to force subcritical masses of enriched uranium-235 or plutonium-239 together to start an uncontrolled self-sustaining chain reaction. The idea is to convert some of the matter in the bomb into energy before the device disassembles itself. In addition to releasing tremendous amounts of energy, these weapons release large amounts of radionuclides (called *fission products*) that constitute the radioactive fallout.

Fusion-type nuclear weapons are more complex. Here a fission-type device is used to trigger a fusion reaction between hydrogen isotopes (usually tritium and deuterium). The fission reaction compresses the fusion material and then superheats it by reflecting the X-rays and gamma rays released from the fission reaction.

It is also possible to coat a nuclear weapon with materials that will increase the amount of radioactive fallout released. This type of weapon might be most attractive to a nuclear terrorist.

Integral to the development of nuclear weapons is the development of weapons delivery systems. These can be relatively simple, like the gravity-triggered bombs detonated over Japan by aircraft, or very complex, such as land-based intercontinental ballistic missiles (ICBMs), which can be in fixed position (in underground silos) or mobile, such as on missile trains. Cruise missiles can be deployed from submarines that can constantly change their position, or from fighter aircraft. Canon-fired nuclear artillery was tested at the Nevada Test Site but never deployed. Deployment of nuclear weapons from space has also been considered but (thankfully) not yet successfully implemented.

Each of these delivery systems has specific purposes. Nuclear weapons dropped from bomber aircraft can be very large, but deployment is delayed because of the plane's relatively slow travel speed (about 15 times slower than an ICBM). A low flight path altitude can make the bomb-carrying plane easy to intercept. Also, with really big bombs, the plane has to distance itself quickly from the detonation site to avoid being blown up or exposing the crew to radiation. ICBMs launched from missile silos travel much more quickly than planes and are more difficult to intercept because of their high altitude. But silos can be targeted by the enemy unless they are "hardened" against such an attack. The likelihood that nuclear weapons deployed by ICBMs will be intercepted can be decreased by loading each warhead with multiple independently targeted reentry vehicles (MIRVs). Submarine-based missiles cannot be easily targeted but must be much smaller than gravity bombs or land-based ICBMs.

Most nations use a combination of these delivery systems to maintain a vigorous strategy of nuclear deterrence, although they may favor one over another.

The health consequences of the development of the current nuclear weapons arsenals are considerable. The aboveground detonations between 1945 and 1963 released substantial amounts of radionuclides into the environment, often referred to as radioactive fallout. Radioactive fallout comprises fission products (including cesium-137, iodine-131, and strontium-90), as well as unfissioned uranium-235 and plutonium-239, vaporized by the tremendous heat of the fireball. These products form a suspension of fine particles about 40 millionths of an inch, or about the size of some viruses. These submicroscopic particles are immediately drawn upward into the stratosphere and spread over the entire hemisphere where the explosion occurs. (Winds of the hemispheres rarely mix, so atmospheric nuclear tests in the northern hemisphere affect only people in the northern hemisphere and contrariwise.)

Most of the fission products released by nuclear bomb testing were short lived and posed few health consequences. However, some were moderate to long lived and have caused, or will cause, important health consequences from exposure to iodine-131, cesium-137, and strontium-90.

A striking example is what happened to sailors on the ironically named *Daigo Fukuryū Maru* (Lucky Dragon) fishing boat, which was just outside the danger zone of the Castle Bravo H-bomb explosion on the Bikini atoll in March 1954. For several hours after the detonation, fallout, including coral that had been made radioactive by the nuclear explosion, rained on the sailors as a white dust. Almost immediately, the crew developed symptoms of acute radiation sickness, and the captain died within seven months. There also was concern about radioactive contamination of fish, especially tuna, similar to what happened and continues after Fukushima. The U.S. government paid $2 million in reparations to Japan because of the Bikini blast.

The United States's H-bomb tests near the Bikini atoll contaminated the Rongelap atoll, whose inhabitants had to be evacuated for several years. Several of them developed thyroid abnormalities, including thyroid cancers. (It was notoriety about the 1946 Operation Crossroads A-bomb test at Bikini atoll that led French engineer Louis Réard to name the swimsuit he invented the "bikini.")

Public concern over radioactive fallout from atmospheric nuclear weapons testing and proliferation led to the 1963 Partial Test Ban Treaty, which banned the tests, and to the 1968 Nuclear Non-Proliferation Treaty, which imposed further restrictions on nuclear technologies. In 1996 many nations signed the Comprehensive Test Ban Treaty, which prohibits the testing of nuclear weapons, thus, at least in theory, preventing the spread of nuclear weapons to compliant nonnuclear states. Unfortunately, several key states (India, Pakistan, and North Korea) have not signed the treaty, and other states have signed but not ratified the treaty, including the United States, Israel, and China. Between 1959 and 2010, the United States entered into at least two dozen treaties, including the Strategic Arms Limitation Treaty (SALT) I and II, the Strategic Arms Reduction Treaty (START I), the Strategic Offensive Reductions Treaty (SORT), the new START, and the Presidential Nuclear Initiatives. Most of these treaties have not been ratified but have helped reduce nuclear arsenals. The U.S.-Russian Megatons to Megawatts program has removed 18,000 warheads from the Russian stockpile and converted highly enriched uranium into material used in U.S. nuclear energy facilities to produce 10 percent of America's electricity. Sadly, this 20-year program will end in 2013.

In 1986 there were 65,000 nuclear warheads worldwide. By March 2012 the number that are operational was reduced to fewer than 4,200, including 1,492 under Russian control and 1,731 under U.S. control. This 95 percent decrease reflects a landmark change in human behavior. Rarely, if ever, in history have adversaries voluntarily given up hugely powerful weapons to such an extent. Perhaps the awesome destructiveness of nuclear weapons underlies this action, or perhaps it is gradual evolution, the moderation of human behavior, or some combination of these factors. Regardless, we are left with the paradox that the development of nuclear weapons of tremendous destructive potential has prompted the largest voluntary disarmament in history.

NEUTRONS IN BOMBS AND IN NUCLEAR REACTOR ACCIDENTS

Neutrons' ability to penetrate deeper into human tissue than other ionizing radiations was not lost on weapons makers. In 1958, in the midst of the Cold War, when densely populated Europe was a potential battleground, the physicist Samuel T. Cohen (1921–2010) at the University of California's Lawrence Radiation Laboratory (now Lawrence Livermore Laboratory) proposed a new type of nuclear weapon, the neutron bomb, or enhanced radiation weapon (ERW). In this device a conventional hydrogen bomb would have its uranium casing removed, allowing more neutrons (with a significantly longer range than charged particles) to escape. They could penetrate even highly shielded buildings or armored tanks with a lethal dose of radiation, because whereas other ionizing radiations, such as protons and alpha particles, can be stopped by lead or other high-density material, neutrons easily penetrate them.

The strength of the blast from a neutron bomb is about half that of a hydrogen bomb, but the amount of radiation released is almost the same—thus a greater fraction of the total energy released—and travels farther; even so, the local blast effect is still in the range of tens or hundreds of kilotons of TNT. But physical destruction is not the point of a neutron bomb. It is meant to guarantee large-scale human death while inflicting less harm on structures. It was designed to kill soldiers who were otherwise protected, and it would be particularly effective against armored tanks. Neutrons would turn their hardened steel protection into a deadly liability by interacting with the uranium and making the tanks radioactive.

The United States, Soviet Union, and France developed neutron devices—the first was successfully tested in 1962 in the United States—but they have not been deployed because of a moratorium on nuclear testing and strong opposition (especially in West Germany, which was a likely battleground) arguing that these bombs made the use of nuclear weapons more likely.

The effect of neutron exposure on humans has been demonstrated in deadly criticality accidents at least seven times. A *criticality* occurs when enough fissionable material (called a *critical mass*) is present for a chain reaction, and all the radiation associated with a fission reaction, including neutrons, is released. The first two criticality events took place at the Los Alamos National Laboratory in 1945 and 1946, when scientists accidentally caused uncontrolled nuclear reactions called a Cherenkov radiation reaction, named for the Russian scientist and 1958 Nobel Prize winner Pavel Alekseyevich Cherenkov (1904–1990), who was the first to do a rigorous study of the phenomenon. Cherenkov radiation is what gives the blue glow to the water around fuel rods in a commercial reactor because of the slowing down of electrons

Neutrons are part of a nuclear reaction and so are present in nuclear facilities. On September 30, 1999, workers at a nuclear fuel–processing plant in Tokaimura, 70 miles north of Tokyo, were moving uranyl nitrate; it is the compound of uranium that results from dissolving the accumulated material on spent nuclear fuel rods in nitric acid. The workers accidentally put more fissionable material into a vat than they should have, and the result was a criticality. Uranium atoms release neutrons at an enormously high speed (called fast neutrons) that is generally not very effective for propagating a chain reaction. But water in the vat into which the uranyl nitrate was placed slowed down the neutrons and caused them to be more effective in sustaining a chain reaction. The criticality continued in a series of pulses, so the radiation ebbed and flowed for about 20 hours and ended only after workers drained all the water from the cooling tank. In all, 667 workers at the plant and nearby residents were exposed to varying amounts of radiation. The three operators who were moving the uranyl nitrate received doses of 3,000, 10,000, and 17,000 mSv, the latter two considered fatal. Seven workers received doses of 5 to 15 mSv, and one nearby resident received more than 20 mSv. Because of his experiences in Chernobyl and Goiânia, Bob was called to Tokyo. He worked with Shigeru Chiba and Kazuhiko Maekawa at Tokyo University and Hideki Kodo and Shigetaka Asano at the Institute of Medical Science to treat the three most severely affected workers.

It takes time and tests to discern whether only gamma radiation has been released or whether neutrons were involved as well. One of the workers reported that he had seen a blue light, suggesting Cherenkov radiation, in which huge amounts of neutrons are released along with gamma rays. (Early nuclear scientists courted danger in assembling a critical mass, approaching criticality without reaching it, which would cause certain death to the experimenter. They called it "tickling the dragon's tail.")

As soon as the worker reported seeing the blue light, the area was evacuated, and the worker was taken to a room near the scene, where he lost consciousness for one minute and vomited within ten. Diarrhea developed within an hour and continued for two days. As there is no explosion or other overt sign of a criticality accident, doctors had to check whether sodium-24 was present in the victim's body to determine whether he had received only a massive dose of gamma radiation or had been bombarded by neutrons as well.

Of the many radionuclides we have in our bodies, sodium-24 is not among them. It is produced when neutrons strike the large amount of nonradioactive sodium-23 (salt) in our body, converting it to radioactive sodium-24, which has a half-life of about 15 hours. To determine whether there was a neutron release from the accident, the worker's urine and sweat were collected and tested; sodium-24 was detected.

Because of the high doses of radiation the victims received, the Japanese medical team and Bob tried to save them by replacing their destroyed bone marrow. They transplanted blood cells from siblings or blood cells obtained from the umbilical cords of unrelated children. Although the transplanted cells effectively replaced bone marrow function in the victims, the exceptionally high doses of radiation to the lungs and gastrointestinal tract caused irreversible, ultimately fatal, damage.

The all-encompassing lethal effects of a neutron bomb are not in question. Nor are the same effects of global thermonuclear war. Immediately after Hiroshima and Nagasaki, people saw that atomic weapons could mean the end of humankind. But according to the late historian Paul S. Boyer (1935–2012), when the Cold War became the norm of international politics and the missile stockpiles of the United States and Soviet Union assured both mass destruction and mutual deterrence, the threat of "instant incineration" somehow became acceptable—or rather, was ignored. The threat from nuclear weapons far outweighs the potential danger of an accident at a nuclear power facility. But perhaps because the large-scale use of nuclear weapons would cause such cataclysmic damage to life on Earth, it is easier to worry about smaller events.

THE EFFECTS OF A "DIRTY BOMB"

National security agencies believe that most state-controlled weapons are out of the reach of terrorists. Instead, they worry that terrorists might try to make their own "atomic bomb" using fissile material that has been stolen or diverted to them. And many of us worry about what might happen if terrorists acquired a small amount of radioactive material and set off a "dirty bomb" (officially known as a "radiological dispersal device," or RDD). A dirty bomb is as explosively destructive as an improvised nuclear device but spreads radioactivity over a small area, injuring some and panicking many. The radiation spread by the loose cesium-137 in Goiânia in 1987 was one of the world's worst radioactive contamination accidents and created many of the effects of a dirty bomb, without an explosion.

Still, the Goiânia incident serves as an indication of what

could happen if terrorists were to set off an RDD in a large city. Around the world about 10,000 radiotherapy units that emit gamma rays from cesium-137 and cobalt-60 are used to treat cancer, especially in developing countries. Radiation therapy centers have on hand a substantial amount of radioactive materials, used for internal radiation therapy (brachytherapy), including cesium-137, cobalt-60, iodine-125 and -131, iridium-192, palladium-103, and ruthenium-106. Cesium-137, cobalt-60, and many other radionuclides are present at hundreds of thousands of industrial sites, universities, and research institutions. Finally, there are radioactive materials at many sites of deactivated nuclear weapons in the United States and ex-Soviet states.

As we saw at Goiânia, someone can steal a radiotherapy machine and remove the radioactive source. Having done this, terrorists could combine it (or several) with a conventional explosive device, which when detonated would contaminate an area of perhaps half a square mile.

Those victims of such a blast who had detectable radiation on their clothes or skin could be quickly decontaminated with thorough washing. The area where the blast occurred would suffer unacceptable radiation exposure, but this could be mitigated by decontamination—washing, shielding, and, if needed, short- or long-term evacuation. But relocating a radioactive source without being detected is not only difficult, it is dangerous, because once the shielding is removed, the radiation levels would likely kill anyone trying to put the material into an improvised bomb. (Ironically, the greatest danger likely would be to the terrorists handling the radioactive material, who would be the most seriously exposed. This, alas, is not a concern for a suicide bomber.)

A small dirty bomb would certainly release radioactive materials, but the greatest danger to people would be not the blast or the radiation but the widespread confusion and hysteria immediately afterward and the subsequent political and economic repercussions. Medical facilities would be overloaded by people with real or, more likely, imagined radiation sickness, and such pandemonium in one city would likely spread to other metropolitan centers, causing additional panic.

A blast using material from a radiation therapy machine or other device that contains a radioactive substance would likely kill or injure people nearby with the percussive force of the explosion or debris propelled by the blast. From a radiation standpoint, the victims would be unlikely to need much immediate medical intervention. Some scientists believe that the crude explosion of a pound and a half of cesium-137 by 4,000 pounds of TNT would so disperse the radioactive material as to render it nonfatal. A scientist with the International Atomic Energy Agency has said, "It's hard to imagine any kind of dirty bomb producing the kinds of mass casualties that we saw on September 11 [2001]."

In fact, terrorists would not need to set off a bomb—they could just spread radioactive material around or smear it on buildings, though they would not then induce the panic factor that an explosion would cause.

None of this implies that the danger from a dirty bomb should be dismissed or the threat not be taken seriously. But bombs aside, the unhappy fact is that radiation spills occur often, and they are cleaned by high-pressure water and special fluids. Plutonium could be cleaned up as well. Most anticipatable radioactive contaminations can be cleaned up if you are willing to spend the time, money, and effort.

Bob and Alexander Baranov wrote in the *Bulletin of the Atomic Scientists* in 2011 that policy makers and the public must be educated about what radiation from such a device can and cannot do. Under almost all scenarios, it is better *not* to evacuate nearby buildings. Rather, people should stay inside (or get inside as fast as possible), close the windows to avoid breathing outside air, shower to wash off as much contamination as possible, and not eat food that could have been exposed to radioactive particles. Evacuation—opening the buildings and decreasing the shielding they offer—usually increases exposure of those near the blast. If there is a radioactive plume, people should stay indoors until it has passed. If there is ground contamination from radioactive material, they should stay indoors until a measurement of the radioactivity is made and an orderly evacuation planned. The point is to avoid panic. People rushing every which way in an attempt to escape will only lead to greater injuries and radiation exposure. As the nuclear weapons expert Bennett Ramberg wrote to us in 2012, in many ways an RDD is more "a weapon of mass distraction than of mass destruction."

NUCLEAR POWER FACILITIES AS RADIOLOGICAL WEAPONS

In *Nuclear Power Plants as Weapons for the Enemy: An Unrecognized Military Peril,* Ramberg describes the possibility of a nuclear power facility being bombed by conventional explosives, allowing for the release of a massive amount of radiation. (In 1981 Israeli military planes destroyed an Iraqi nuclear reactor, but it was under construction and had no nuclear fuel and thus released no radiation.) He details how prior to the attacks on September 11, 2001, governments showed too little concern over this vulnerability, not only for nuclear but for chemical facilities, and quotes a report of Great Britain's Royal Commission on Environmental Pollution: "The vast increase in the chemical process industry over the last few decades has created many industrial plants where the consequences of damage from armed attack could be extremely serious. The unique aspect of nuclear installations is that the effects of the radioactive contamination that could be caused are so long lasting. If nuclear power could have been developed earlier, and had it been in widespread use at the time of [World War II], it is likely that some areas of central Europe would still be uninhabitable because of ground contamination by caesium."

A 2005 report by the U.S. Congressional Research Service notes that nuclear power plants "were designed to withstand hurricanes, earthquakes, and other extreme events, but attacks by large airliners loaded with fuel, such as those that crashed into the World Trade Center and the Pentagon, were not contemplated when design requirements were determined."

The nuclear industry's response is that even if there were such an attack, penetration of the reactor vessel holding the nuclear fuel is unlikely, and that a "sustained fire, such as that which melted the structures of the World Trade Center buildings, would be impossible unless an attacking plane penetrated the containment completely, including its fuel-bearing wings."

A terrorist assault on a nuclear power facility could induce a meltdown and trigger a release of radiation. To help thwart this possibility, an April 2003 order from the U.S. Nuclear Regulatory Commission (NRC) states that preparation from an attack must "represent the largest possible threat against which a regulated private guard force should be expected to defend under existing law." Nuclear facilities regularly conduct exercises in which personnel respond to multiple attack scenarios. The vulnerability of nuclear power facilities is a continuing concern, although in the decade since the 9/11 attacks, security has improved.

RADIATION ACCIDENTS AND MISADVENTURES

To focus solely on terrorism as the means to spread radiation is to miss the larger point that radioactive material exists all over the world and that it is used, stored, carried, and sometimes misplaced and forgotten about by human beings. Humans are guaranteed to make errors. The Goiânia story is not an isolated example of highly radioactive material being mishandled. In late 1983, in Juárez, Mexico, across the border from El Paso, Texas, there was an incident remarkably similar to that in Goiânia. An electrician, unaware of the danger, opened a discarded radiation therapy machine capsule filled with cobalt-60 he had collected in his pickup truck, and in driving to his junkyard scattered radioactive pellets on the road.

The bulk of the cobalt-60 pellets were mixed with scrap and taken to two foundries, where the metal was made into table legs and reinforcing rods for construction. In all, thousands of tons of metal were contaminated. The radioactive metal went undetected until a truck carrying a load of it in New Mexico made a wrong turn near the Los Alamos National Laboratory, tripping a radiation alarm. Officials tracked down the remaining objects in several states and in Canada, as well as rebar used in hundreds of new homes built in at least four Mexican states.

The accident released about 100 times the radiation that escaped from the Three Mile Island nuclear power facility and exposed more than 200 people to low but significant doses of radiation over a long period. It is considered perhaps the worst spill of radioactive material in North America.

Stories such as these occur with alarming frequency. In 1998 the International Atomic Energy Agency organized the first conference devoted to the safety of radiation sources and the security of radioactive materials. The conference was cosponsored by the European Commission, the International Criminal Police Organization (Interpol), the World Customs Organization, and the French Atomic Energy Commission. The reports showed how often radioactive material is mishandled, lost, or simply overlooked.

The NRC is notified of about two hundred lost or stolen radioactive sources each year. Since 1983, twenty of them have been melted at steel mills or other foundries and recycled into new metal. The NRC considers these instances to be only a small fraction of the devices that have been improperly recycled, even though many scrap dealers and metal recycling plants have radiation-detecting equipment, often at their front gates. Nonetheless, small pieces of radioactive material can sometimes pass through several detection checkpoints before being discovered. Because of the high cost of detection equipment, only about half of all scrap dealers in the United Kingdom have installed it or at the least possess hand held detectors to monitor their inventory. (The problem is likely considerably greater in less developed countries.)

In 1998, at an iron smelting plant near Cádiz, Spain, a medical machine containing cesium-137 passed through monitoring equipment undetected and was melted along with all the other products in the smelting process. The resulting gases were released through the plant's chimney (equipped with radiation detectors, which were not working) and dispersed into the atmosphere. Temporary radiation measurements of 1,000 times normal were detected by monitoring devices in France, Switzerland, Italy, Austria, and Germany. Between 1982 and 1984, scavenged radioactive metal was melted into rebar and used to construct about 2,000 apartment units in northern Taiwan. One report states at least 10,000 people received long-term, low-level radiation and several died. An analysis suggested an increased cancer risk for residents. These data are used to argue that low doses of radiation exposure over a long interval can cause cancer.

In Russia, Bob experienced firsthand the consequences of an errant radiography machine. Modular units used to construct buildings are stacked one atop the other and then are subjected to a nondestructive testing technique called radiography, to ensure the steel is uniform in strength and that the joints are proper. Field radiography generally uses a strong radioactive source that emits gamma rays; it is placed on one side of an object, and photographic film is placed on the other side. The process is akin to taking an X-ray, but it is done with radioactive materials rather than a radiation-producing device such as an X-ray machine. One typical source of radioactivity in a radiography machine is iridium-192 (which has a half-life of about 74 days). These machines are purposely designed to be portable and so might easily be stolen to construct an RDD. (More radiation accidents can be attributed to industrial radiography than to any other use of radioactive material.)

During the construction of a prefabricated apartment building in Kharkov in the 1980s, a machine that used cesium-137 to examine the units detached from its tether and, somehow unnoticed, was incorporated into a concrete slab that became a bedroom wall of an apartment. Families who lived in the apartment became ill and moved out. Others took their place but also fell ill. Brothers who shared the bedroom and slept with their feet to the wall with the radioactive source in it developed skin rashes and ulcers on their legs and eventually bone marrow failure. The older brother, who also developed a bone cancer (osteogenic sarcoma) of his ankle, died. Residents and local authorities blamed these health problems on bad luck. Eventually Bob's Russian colleague Alexander Baranov heard the story. He immediately understood that the cause might be radiation poisoning. Inspectors were sent to the apartment and quickly discovered and recovered the radioactive source embedded in the wall. One child and his mother were promptly hospitalized and treated.

CHAPTER 8

NUCLEAR POWER AND RADIOACTIVE WASTE

IS THERE A SAFE WAY TO GENERATE ELECTRICITY?

Electricity has fundamentally changed and improved society, industry, the economy, and human well-being. In a 2009 article in *The Lancet,* Anil Markandya and Paul Wilkinson detailed the results of a study of electricity generation and health. Between 1820 and 2002, annual real per capita income in western European countries rose from about $1,200 to $19,000, a sixteenfold increase. In the same period, average life expectancy rose from 40 to nearly 80 years. The change from animal to steam power and then to gas in the nineteenth century improved productivity and vastly decreased the health problems associated with animal waste. The widespread use of electricity from 1900 onward has enabled an exponential rise in industrial productivity that, coupled with rapid technological advancement, has been a boon to standards of living and human health. Not having to rely on wood to heat a home and cook meals or on candles for light has reduced the risk of fires, cleaned the air in rooms, and made houses warmer in cold weather, all of which (along with antibiotics and other medical advances) improved health and lengthened lifespan. Additionally, the use of electricity has become remarkably efficient. One unit of electric energy produces four and a half times more output than one unit did in 1850.

Unfortunately, every means of electrical generation in great quantity comes with substantial drawbacks, including health risks from pollution and/or storage of the by-products of that process. The production of natural gas (a fossil fuel) harms aquifers, burning natural gas releases carbon dioxide, and gas leaks release methane, a greenhouse gas that, ton for ton, traps 25 times more heat than carbon dioxide. Coal, the most widespread source of power, releases at least 84 hazardous air pollutants and toxins, including carbon dioxide; acid gases such as hydrogen chloride; toxins such as mercury, arsenic, lead, benzene, and formaldehyde; and radionuclides, including thorium and uranium. But even those means that send no emissions into the atmosphere are problematic. Wind and solar power depend on the weather. Wind turbines kill birds and bats, and some people think they scar the landscape. Conventional solar energy facilities require large tracts of land. Mining copper for use in pipes brings to the Earth's surface radioactive thorium, radium, and uranium, which are found with the copper. Some data suggest that the radiation dose from conventional solar energy per megawatt may exceed that of nuclear energy. Photovoltaic electrical production uses highly toxic materials that, unlike radionuclides, never decay and remain a health hazard forever. Hydroelectric power, which requires dams and reservoirs, affects fish and birds and alters rivers. Spawning salmon die trying to return to their homes. And many of the most feasible sites for damming are already used.

Though each of the renewable technologies makes important contributions to overall electricity generation, none is likely ever to replace a large fraction of the energy now created by burning fossil fuels and running nuclear power facilities. Even the most seemingly environmentally friendly project, a hydroelectric dam to prevent flooding and generate electricity, can have unintended consequences. The Aswan High Dam on the lower Nile River in Egypt prevents annual flooding and produces a substantial proportion of Egypt's electricity. It slowed the flow of the upper Nile and the seasonal fluctuations in water level in canals beside the river. But the slowed water flow resulted in increased infections, with chlamydia and schistosoma causing 1 million cases of blindness and schistosomiasis (severe swelling of the lower extremities).

Every year the single greatest transfer of wealth in the history of civilization is repeated in the purchase of fossil fuels, particularly oil. That figure will only grow in the foreseeable future. The increase in worldwide population and economic growth in developing nations will bring rising demand for all forms of energy. As the standard of living rises for the people in those countries, many more billions want electricity for their homes and the myriad powered amenities of life. They will also want gasoline for their cars.

Americans use more than 10 billion kilowatt-hours of electricity a day, more than twice the amount per capita of any other nation. (Using a 100-watt lightbulb for ten hours consumes one kilowatt.) About 45 percent of U.S. electricity is generated by coal. (The five largest coal users—China, the United States, India, Russia, and Japan—account for 77 percent of global coal use.) The United States' 600 or so coal-burning plants produce at least 500 megawatts (500 million watts) a day, each with enough to power as many as 250,000 homes. Each of those 600 plants each burns on average 1.4 million tons of coal per year. According to the Union of Concerned Scientists, each year every one of those plants emits into the air:

- 3,700,000 tons of carbon dioxide (CO_2), the primary man-made contributor to global warming. It is the equivalent to cutting down 161 million trees, which absorb carbon dioxide while they are alive.
- 10,000 tons of sulfur dioxide, which causes acid rain that damages forests, lakes, and buildings and forms small airborne particles that can penetrate deep into our lungs.
- 500 tons of small airborne particles, which cause chronic bronchitis, aggravate asthma, and cause premature death, as well as haze that obstructs visibility.
- 10,200 tons of nitric oxide, the amount emitted by half a million late-model cars; nitric oxide leads to formation of smog (ozone), which inflames the lungs, burns through lung tissue, and makes people more susceptible to respiratory illness.
- 720 tons of carbon monoxide, which causes headaches and places additional stress on people with heart disease.
- 220 tons of hydrocarbons, volatile organic compounds that form ozone.
- 170 pounds of mercury. Just one-seventieth of a teaspoon deposited on a 25-acre lake can make the fish in them unsafe to eat.
- 114 pounds of lead, 4 pounds of cadmium, other toxic heavy metals, and trace amounts of uranium.
- 225 pounds of arsenic, which will cause cancer in 1 out of 100 people who drink water containing 50 parts per billion of this compound.

Waste created by a typical 500-megawatt coal plant each year includes more than 125,000 tons of ash and 193,000 tons of sludge from smokestack scrubbers. Nationally, more than 75 percent of this waste is disposed of in unlined, unmonitored onsite landfills and surface impoundments.

Toxic substances in the waste, including arsenic, mercury, chromium, and cadmium, can contaminate drinking water supplies and damage human organs and the nervous system. One study found that 1 out of every 100 children who drank groundwater contaminated with arsenic from coal power plant wastes was at higher risk of developing cancer. Ecosystems too have been damaged—sometimes severely or permanently—by the disposal of coal plant waste.

The 2.2 billion gallons of water that are cycled through the coal-fired power plant are released back into lakes, rivers, and oceans. This water is hotter (by as much as 25 degrees Fahrenheit) than the water it enters. This thermal pollution can decrease fertility and increase heart rate in fish. Typically, power plants also add chlorine or other toxic chemicals to their cooling water to decrease algae growth. These chemicals are also discharged back into the environment. About two-thirds of the heat produced from burning coal to produce electricity is released into the atmosphere or the cooling water.

Markandya and Wilkinson studied the various means of producing electricity through every stage of the cycle to the point where they cause harm in humans. Among their findings:

Coal:

- Up to 12 percent of coal miners develop at least one of five potentially fatal lung diseases.

- Sulfur dioxide and nitrogen oxides, released in the generation of electricity from coal, contribute to the creation of secondary particles that are harmful to lungs.
- Worldwide, thousands of coal miners die annually, most of them in China. The U.S. Department of Labor reports that in 2006 to 2007, 69 American coal miners died in mines and 11,800 were injured.

Oil and gas:

- The primary and secondary particles released into the air are much smaller than those from coal, and therefore the health effects are exponentially less. Health problems from gas are half those from oil, but those associated with coal are ten times greater.

The environmental dangers of oil drilling and transport accidents and the vast expense of cleanup are evident in the 2010 *Deepwater Horizon* accident in the Gulf of Mexico and the 1989 grounding of the *Exxon Valdez* on Bligh Reef in Prince William Sound in Alaska. These are only the two best known of many spills, explosions, and other accidents.

But the production of electricity is not the greatest cause of pollution. Cars, trucks, and other forms of transportation that use oil refined into gasoline are the source of about one-quarter of the hydrocarbons and greenhouse gases in the atmosphere, more than one-third of the nitrogen oxides, and more than half the carbon monoxide—a colorless, odorless, and poisonous gas that blocks the transport of oxygen to the brain, heart, and other organs. Fetuses, newborns, and people with chronic illness are most susceptible to it.

The health risks from fossil fuels, as serious as they are, are overshadowed by the threat that their emissions pose to our biosphere, which provides humanity with fundamental life support, not least by shielding us from cosmic radiation. As the atmosphere degenerates, the ozone layer depletes, and global warming exacerbates these changes, the effect on life as we know it may be catastrophic. Using fossil fuels could result in more radiation-related cancers (consider, for example, skin cancers from UV radiations) than using nuclear energy.

There are approximately 435 nuclear reactors in the world. A normally operating nuclear power plant emits very small amounts of radioactive gases and liquids, as well as small amounts of radiation. People living within 50 miles of a nuclear power plant receive an average radiation dose of about 0.0001 mSv per year, or one-thirty-three-thousandth of what the average American receives annually from natural background sources of radiation. Those who live in proximity to coal plants, however, receive substantially more radiation. Coal, before it is burned, contains only trace amounts of uranium and thorium that pose no danger. But burning coal removes the carbon and impurities and leaves "fly ash," in which uranium and thorium are concentrated up to ten times their pre-burn level. The natural radioactivity in coal is released by burning and is redistributed into the atmosphere. Fly ash carries as much as 100 times more radioactivity into the surrounding environment than a nuclear power plant producing the same amount of energy. Fly ash also seeps into the water table and can be absorbed by crops we eat. Only a small fraction of fly ash is recycled into products such as concrete.

Most of the radioactivity from nuclear plants is man-made—something that never existed until we meddled with

Mother Nature. But the facilities capture most of the radioactivity they produce using filters. Some radioactive gases and liquids are released into the environment—usually, but not always, under controlled conditions so that they dissipate into the atmosphere or are diluted by water. Most of an operating nuclear power plant's direct radiation is blocked by the steel and concrete of the building and containment structures. The remainder safely dissipates in the controlled, uninhabited space around the plant and is not a danger to the public.

Granite and some other stones are thousands or even millions of years old. As such, they contain radionuclides in their matrix, including thorium-232, uranium-238, and radium-226. As these radionuclides decay, they release radon-222. The radiation levels in New York's Grand Central Terminal or in front of the U.S. Capitol are higher than the background level permitted in a nuclear power facility. However, nuclear power facilities are not without their hazards, as we have seen at Chernobyl and Fukushima Daiichi.

The typical nuclear power reactor will generate 24 tons of waste in a year. For all reactors worldwide combined, this amounts to an annual total of about 10,400 tons—about 416,000 tons in the last four decades. The material with the highest level of radioactive waste is spent fuel rods. These are typically stored in water pools on the reactor sites, although some are in concrete and steel dry casks. This sounds threatening. Storage of nuclear waste is, of course, a public concern, in part because it is radioactive for millennia and in part because anything with the word "nuclear" in it is scary. Fears arise mainly because of concerns about the testing and use of nuclear weapons, accidents like Chernobyl and Fukushima, the threat of global annihilation, and exaggerations of the possible consequences of nuclear accidents. Many of these fears have been increased by inadequate education and misinformation.

It is natural to worry about what will happen if radioactivity from stored fuel rods leaks from its containers, seeps into the ground, and enters the water supply. So for a moment let's be counterintuitive and compare waste from coal plants with waste from nuclear power facilities. (Put aside the pollution that results from coal mining.) The burning of coal produces about 130 million tons of ash and other mostly toxic waste annually in the United States; that is 12 million times the weight of waste produced by nuclear facilities worldwide.

It is estimated that if the entire U.S. electricity use were met by nuclear reactors, the high-level radioactive waste volume from 350 years of production would fit into a cube measuring 200 feet on a side. When coal ash is sealed in dry, lined storage pits in the ground, it is thought to be safe, but often it ends up being dumped in huge ponds or stored in the open air. In 2009 in Kingston, Tennessee, an enormous pool of coal waste that had grown over half a century burst through its barriers and sent approximately 1 billion gallons of hazardous sludge into the Emory River, a toxic spill 100 times greater than the leak from the *Exxon Valdez*.

Then there is the pollution from coal and all fossil fuels that enters the atmosphere. We worry about storing nuclear waste on Earth, but 350 years of that waste would often be smaller than the waste pile generated by one coal-fired station.

It would be wonderful if there were a form—or several forms—of energy production that could meet the world's energy needs and did not have the long-term dangers of nuclear power or the immediate dangers of fossil fuels. In the future, nuclear reactors designed to operate at very high temperatures may be able to split water molecules to produce hydrogen naturally and create energy by fusion, with very little release of carbon dioxide. But we are not there yet and may not be for a long time. (A standard joke among nuclear physicists is that fusion is always 50 years in the future.)

Contributing to the global problem of rising levels of carbon dioxide in the atmosphere is deforestation in developing countries. Trees are felled so that the wood can be burned for heating and cooking. Efficient nuclear energy could replace that. Nuclear technology has progressed considerably in recent decades. Advanced, fast neutron reactors, combined with the reprocessing of nuclear fuels that lead to the consumption of more fissile material, require less uranium ore mining and milling (thus cause less environmental damage) and produce better systems for waste disposal. As a result of the U.S. government–imposed policy of not reprocessing spent fuel, the fuel from light-water reactors still contains more than 90 percent of the potential energy needed for nuclear fission and could be efficiently recycled under appropriate circumstances.

Nuclear energy also has a tremendous advantage in energy density. Burning 1 kilogram of coal can power a 100-watt lightbulb for about 4 days; a kilo of natural gas can keep it going for about 6 days. But a kilogram of uranium in a light-water reactor can power the lightbulb for 140 years.

Whether you think burning fossil fuel or using nuclear energy is the more dangerous option may be influenced by whether you believe that global warming is real and man-made. If you believe that global warming is real and that emissions of fossil fuels, chlorofluorocarbons, and other greenhouse gases are making it steadily worse with potentially dire consequences, then contrary to what you might have long thought, nuclear power may be the lesser danger. It is surprising how few environmentalists favor nuclear energy. (One self-described "tree hugger" whose research led him to do so is F. Ward Whicker of Colorado State University, whose work over the past several decades has focused on how to clean up radioactive waste.)

To put it another way, we fret about what might happen with sealed nuclear waste kept far from a population center in an area of a few hundred acres. If everything imaginable went wrong, including a meltdown at a nuclear power facility, radioactivity would render an area uninhabitable and enter the water supply, affecting many thousands or perhaps a few million people. That would be terrible. (Using Chernobyl as a guide gives a rough idea of what might be expected. So far, as we have mentioned, there are about 6,000 thyroid cancers [exclusively in children], a possible slight increase in leukemia, and no convincing increase of any other cancer in 25 years since the accident. The thyroid cancer increase in unfortunate but largely preventable, and there are only 15 deaths so far. Estimates of possible cancers over 80 years from the accident range from 11,000 to 16,000, but they could be as low as 6,000 or as high as 25,000 or more. Over those same 80 years, there likely will be more than 100 million cases of cancer in the population of the former Soviet Union unrelated to the Chernobyl accident. An individual's increased chance of developing cancer is about one-half of 1 percent above the cancer rate for the general population, about 38 percent for women, and about 45 percent for men.)

Yet we seem not to care that by releasing ever-increasing amounts of ozone-depleting contaminants from fossil fuels and greenhouse gases into the Earth's atmosphere for more than a century, we have created the largest waste dump in our solar system, and now those pollutants threaten human life as we know it.

Deaths per Terawatt-hour (1 Trillion Watts)

	Accidents	Air Pollution
Coal	0.02	25
Gas	0.02	3
Oil	0.03	18
Nuclear	0.003	0.05

RADIOACTIVE WASTE

Nuclear fuel consists of half-inch-long ceramic pellets of uranium encased in 12-foot-long rods resembling spears. Most of the fuel is uranium-238, which is not ordinarily fissionable, but adding 3 to 5 percent uranium-235 by weight enriches it sufficiently for a chain reaction. (Sometimes plutonium-239 is used as the fissile component.) At the end of 2009, the United States had about 140 million pounds of spent nuclear fuel from nuclear power facilities. The amount is growing by about 2,000 tons annually. France, the United Kingdom, and all countries that use nuclear power also have large inventories of spent nuclear fuel.

Light-water reactors, the most common reactors in the United States and western Europe, typically use fuel that consists of a large number of ceramic pellets inserted into zirconium alloy tubes (fuel rods), which are then bundled together into fuel assemblies. (Zirconium is used because it absorbs relatively few neutrons released by the fissioning of the uranium-235, and it also resists corrosion.) Space between the ceramic pellets and the zirconium cladding is filled with helium gas to improve the conduction of heat generated from fissioning the uranium to the cladding. There are about 200 fuel rods per fuel bundle and about 150 fuel bundles are in the reactor core. To control the numbers and speed of neutrons passing through the fuel, control rods are inserted through the top between the fuel assemblies. After some fraction of the uranium fuel has fissioned ("burned") to produce hot water and steam for power production, the fuel rods are referred to as "used" or "spent."

Every nuclear power facility stores its spent fuel rods in 40-foot-deep steel-lined pools that in many cases are surrounded by several-feet-thick reinforced concrete. The water in the pools is kept at a low temperature to cool the rods, which still release heat from residual radioactivity, and shield the environment and workers from this radiation. "Spent nuclear fuel" generally refers to fuel rods that can no longer sustain a nuclear reaction at reasonable cost but that still contain significant quantities of unfissioned fuel. These spent fuel rods continue to generate heat through radioactive decay of the almost 200 radionuclides in the fuel, including unfissioned uranium-235 and plutonium-239 created in the reactor. There are built-in cooling safeguards—if cooling were lost, there would be sufficient time to restart the pumps before the water covering the fuel elements boiled off. Plants have back-up power-generating and water-pumping equipment in case a fire, explosion, or other catastrophe renders the normal pumping equipment inoperable.

Although the process of producing electricity using nuclear energy may seem complex and potentially dangerous, the risk of death at a nuclear power facility is about ten times less per unit of electricity produced than at a conventional coal-fired electricity-generating station. Sadly, no system is foolproof. The tsunami at Fukushima Daiichi knocked out all of the cooling systems, not only to the nuclear reactors but also to the cooling pools containing the used nuclear fuel. Emergency workers were forced to use firefighting equipment to pump seawater into the pools. Although this solved the problem of maintaining an acceptable temperature in the pools, salt and other chemicals in the seawater reacted with the zirconium cladding on the fuel rods, causing them to deteriorate. Japanese technicians are now trying to understand the interaction of the seawater in the spent fuel rods and a large volume of radioactivity-contaminated water. Except at the Chernobyl meltdown in 1986, no one has died as a result of radioactivity released by a commercial nuclear power facility. Still, potentially, waste of this kind is extremely dangerous. The nuclear industry acknowledges this and says it manages the production of energy and the storage of waste with great care. It is imperative that regulators ensure this.

About 80 percent of U.S. spent fuel rods are in the pools at the reactor in which they were used. The remainder are stored in dry casks, huge steel structures about 15 feet high, that stand in secured fields at protected sites. Spent fuel rods are typically kept in water pools for five years before there has been sufficient cooling to allow transfer to dry cask storage. Dry storage containers are surrounded by additional steel, concrete, or other materials that shield the environment and the public from the residual radiation of the spent fuel rods. In November 2010 there were 63 so-called "independent spent fuel storage installations" at 57 sites in 33 states, with more than 1,400 casks in all. In addition, the United States has 10 decommissioned nuclear power reactors at 9 sites that, although they perform no nuclear operations, hold approximately 3,100 tons of spent nuclear fuel. However, these sites cannot be fully decommissioned until this material is moved to a consolidated storage or disposal facility.

What to do with used nuclear fuel is a continuing debate

and a political and public policy quandary. In 1987 the Department of Energy was ordered to create storage for the radioactive waste from the nation's then 104 nuclear reactors by 1998. A decision was made to bury it in a specially engineered repository deep inside Yucca Mountain, Nevada, a mass of volcanic rock 90 miles from Las Vegas that was part of the Nevada Test Site for nuclear weapons. Though the U.S. government spent the next two decades readying Yucca Mountain, some Nevada citizens and their congressional representatives have resisted having radioactive waste entombed there. Meanwhile, a growing amount of radioactive waste stays in ever-diminishing space at the nuclear facilities that used it, in pools not designed for long-term storage. Reactors that go online in 2015 and beyond are mandated to have storage for at least 18 years' accumulation of spent nuclear fuel.

Dangerous as the spent fuel is, there is surprisingly little of it, which makes its safe storage practical and of course imperative. If all the ceramic uranium fuel pellets produced and used by U.S. nuclear power facilities over the past 50 years were assembled in one place, they would cover an area about the size of a football field, to a depth of about twenty-one feet. (This should be done only in theory; actually stacking them like that might set off a chain reaction.) The longevity of nuclear waste—plutonium-239 has a half-life of 24,000 years, so it takes 240,000 years for most of it to decay—seems a natural and major concern, but it lacks context. Many other toxic wastes and chemicals, such as lead from batteries, don't decay at all—they have infinitely long half-lives. Just because something exists does not mean we will be exposed to it. *All* extremely hazardous waste needs to be quarantined, and a vast amount of it is not.

Scientists at the Sandia National Laboratories near Albuquerque, New Mexico, have looked at the possibility of storing nuclear waste 12,000 feet below the ocean's surface in soft, gooey mud. One suggestion is to load the waste into torpedoes that would penetrate a few hundred feet into the seabed. If the casing around the waste were ever to corrode or leak, a tiny fraction might release, but the rest would be tied up with the tectonic plate sediments, being buried deeper and deeper with time. Other scientists have suggested sending spent nuclear fuel into space, despite the attendant risk of rocket failure.

As the impasse in Nevada continues, the Waste Isolation Pilot Plant for storage of high-level nuclear waste near Carlsbad, New Mexico, has been built in an old salt mine in a huge salt formation. The 2-mile-long tunnel is deep underground in dry salt beds that for aeons have been impervious to water. The facility is meant to last 10,000 years, during which time humans presumably will have found more effective ways of dealing with these by-products that can stay radioactive for nearly a quarter of a million years. As the uranium in spent fuel still retains about 95 percent of its potential power, reprocessing and recycling it is a popular though currently very expensive option in several countries but not in the United States, where it is unpopular as well as illegal. Many scientists consider the Carlsbad site preferable to Yucca Mountain because the absence of groundwater means the salt beds will not be leached away and will always be dry—unless the whole Earth climate changes, at which point this will be the least of our worries.

What if radioactive waste could be recycled into virtually limitless fuel for a reactor? It could be done. Fast breeder reactors, once they are fully charged with plutonium fuel and material that can convert to fissile fuel by absorbing neutrons, generate more fuel than they use. Extra neutrons created by fission can be used to convert a nonfuel (uranium-238) into a fuel (plutonium-239). Moreover, once a fast breeder reactor is operational, it requires only natural uranium-238, rather than uranium-238 enriched with uranium-235, for fuel.

Along with these advantages, however, fast breeder reactors have many drawbacks. They are far more complicated and expensive to build and operate than normal fission reactors. Of greater importance, the plutonium in a fast breeder reactor can be removed and used by terrorists or rogue states to make nuclear weapons. Also, to extract the plutonium, the fuel has to be reprocessed, which creates radioactive waste and potentially high radiation exposures. In 1977, the Carter administration halted spent fuel reprocessing and that has been the official U.S. government policy ever since. In 1967 France attempted to commercialize fast breeders but stopped in 1983 after technical problems coupled with commercial failure made it impractical. An attempt to reprocess spent fuel through transmutation was halted in 2009. India, Japan, and the former Soviet states are among the countries with fast breeder reactors, none of them commercial.

In the 1960s an earlier version, called simply breeder reactors, were considered the better choice because uranium was very expensive at the time, and the cost of enriching natural uranium with the uranium-235 isotope was also very expensive. (Enrichment to rare uranium-235 is expensive; uranium-238 is plentiful and relatively inexpensive.) As the cost of enriching uranium has decreased significantly and enrichment has been made easier through new technology, commercial considerations have made breeder reactors less attractive. A possibly economically-feasible fast breeder reactor is being built by Russia with U.S. input. Operation is to begin in 2014.

Another possibility for lessening high-level nuclear waste is mixed oxide fuel (MOX). MOX is a blend of plutonium and uranium oxides (natural uranium-238, reprocessed uranium, or even depleted uranium). A benefit of MOX is that it uses

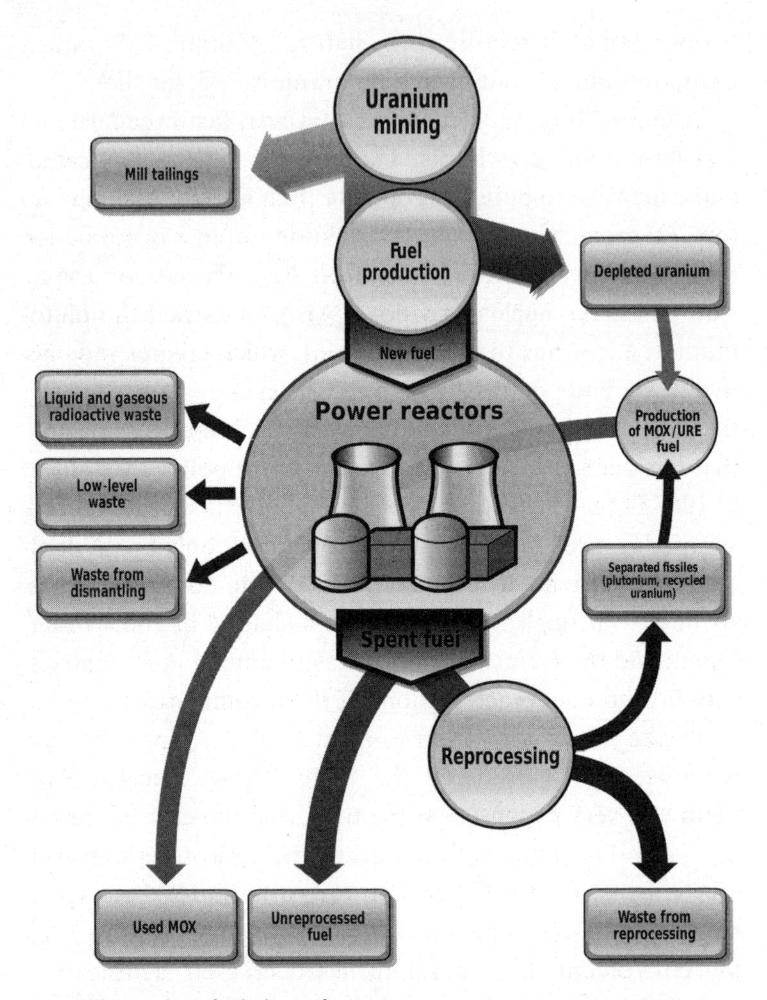

The nuclear fuel chain, from mining to use to separation into high- and low-level waste, and the possible ways to reuse waste.

the weapons-grade plutonium produced for nuclear weapons, so countries would no longer have to store it and protect it against theft. Tens of thousands of nuclear weapons have been retired from the U.S. inventory, each one a significant source of plutonium that could be used as fuel. A drawback is that widespread use of MOX for nuclear reprocessing would increase the risk of nuclear proliferation. The plutonium-239 that is separated might be more readily obtained by terrorists.

High-level radioactive waste, such as nuclear fuel rods and waste tanks left over from making plutonium for nuclear weapons, exists in former plants at Hanford, Washington; Rocky Flats, Colorado; and the Savannah River Site in South Carolina. They need to be treated very carefully and quarantined (as at Carlsbad) so they do not endanger people or the environment. Scientists have the technical ability to do this. It is the political will that is lacking.

Another sort of radioactive material is low-level radioactive waste—a much lower form than spent reactor fuel or materials used in processing it. In the United States this waste is at more than 100 sites in thirty states, covering about 2 million acres. Cleanup generally involves excavation, transport and disposal of soil, and pumping and treating of groundwater. F. Ward Whicker has written that more than $60 billion has already been spent without a corresponding reduction in risk; in some cases, these activities simply move contaminated material around and spread it through the air and water. The safest, most economical and practical way to treat such waste sites is to keep people away from them and monitor the level of radioactivity. Studies by environmental scientists have found cesium-137 and strontium-90 in some of the reservoirs used for cooling reactors and in their sediment, as well as in local fish and birds. But the levels are low enough that according to current knowledge, they have no health impact. Regulations for cleanup, often based on emotion and political expedience rather than on science, require digging up soil with levels of radiation that are similar to natural background levels. It is impossible to clean up every trace of radioactive material because the soil and rock we live on is naturally radioactive. Just moving dirt may cause ecological damage without any benefit.

Between 1946 and 1972, the U.S. government disposed of low-level nuclear waste by shallow land burial at government sites or by ocean dumping in areas approved by the Atomic Energy Commission (the forerunner to the Nuclear Regulatory Commission). It was done without much foresight or care until 1972, when the London Dumping Convention (formally the Convention on the Prevention of Marine Pollution by Dumping of Wastes and Other Matter) was signed; it became effective in 1975. As of 2005, eighty-one countries were signatories. Prior to that agreement, low-level radioactive waste was carefully transported but, as a 1980 EPA report put it, because it was "regarded primarily as garbage, precise reports were not kept of the disposal operations." Items disposed of included equipment, lab clothes, tools, and other matter contaminated by exposure to radioactive materials. The loads of waste also comprised "radioactive cobalt, strontium, americium, and cesium. At times, they also may include small quantities of 'source materials,' such as uranium and thorium, or traces of 'special nuclear materials' such as plutonium or enriched uranium."

The waters around the Farallon Islands, about 25 miles offshore of San Francisco, received approximately 99 percent of the radioactive material dumped by the United States into the Pacific Ocean. About 98 percent of the waste dumped off the East Coast was at the Atlantic 2800 Meter Site, 120 miles offshore from New York City, at a depth of 9,200 feet (2,800 meters). It is being slowly covered by sediment (one hopes). Ocean sites used by other nuclear nations are offshore of New Zealand, western Europe, North Africa, China, and Japan. In addition, the former Soviet Union scuttled atomic submarines under the polar ice cap. From 1946 through 1962, the United States dumped approximately 89,400 containers with an estimated radioactivity of more than 3 terabecquerels (3 trillion becquerels; a submarine's reactor would have about half as many becquerels of radioactivity). Ocean dumping began to be phased out in 1963, and in 1970 ocean dumping ended.

The amount dumped was a tiny fraction of the natural radioactivity of the oceans, which comes mostly from potassium-40, but also from uranium, tritium, carbon-14, and rubidium-87. The total amount of natural radioactivity in the sea is estimated to be 14 zetabecquerels (ZBq)—1 trillion billion becquerels. This includes the approximately 4.5 billion tons of uranium in the world's oceans, enough to power every nuclear power facility on Earth for the next 6,500 years. Researchers at the U.S. Department of Energy's Oak Ridge National Laboratory and Pacific Northwest National Laboratory announced in August 2012 that using technology developed in Japan, they have more than doubled the amount of uranium that can be extracted from sea water.

Although there was, and continues to be, considerable debate and concern over the 80 percent of atmospheric releases from Fukushima Daiichi that were deposited in the ocean, most people are unaware that for thirty years many countries intentionally dumped radioactive wastes into the seas.

Dumping radioactive materials into the oceans seems, at first, to be a terribly bad idea, and it probably is. But from a safety standpoint, the huge dilution of these radioactive materials in the vast volume of ocean water has the impact of reducing the potential damage to living creatures, should there be a leak in the containers that house the material. For example, much of the cesium-137 released from Fukushima Daiichi was diluted in seawater. Because the bodies of living creatures perceive cesium-137 as similar to potassium, to be ingested each radioactive atom of cesium has to compete with an infinitely greater amount of potassium. Dilution and competition make it extremely unlikely that the cesium-137 would have a long-lasting effect on fish or other sea life. (Bottom-feeding [demersal] fish may be an exception.) The protective effect of dilution by the oceans or by running water (rivers) is one reason nuclear power facilities are sited near water. (Cooling requirements are another consideration.)

This protective effect does not apply to closed bodies of freshwater. For example, some freshwater lakes in Scandinavia were contaminated by cesium-137 released from Chernobyl. Fish from these lakes may contain cesium-137 levels exceeding regulatory limits for safe consumption.

Soon after the massive release of radioactivity at Chernobyl, pine trees in the vicinity died, creating a "red forest." This is of particular interest, as some kinds of pine trees have a susceptibility to radiation similar to that of humans. But the environment recovered, unexpectedly quickly: today much of the wilderness around the former plant is thriving, in part because of the absence of humans. Of course, this is not to say there were not important consequences from the released radiation. There still are areas where animals are sterile. But there are also large populations of deer, mice, and birds; wildlife from other areas is repopulating the area, and radiation levels are declining, albeit slowly.

None of this should minimize the extraordinary costs of cleaning up a radiation accident, work that can last decades. No other kind of energy accident can compare in long-term effort. But the costs of cleaning up the contamination of the many combined fossil fuel spills and toxic waste dumps is also high, and the dangers are also long lasting.

Radioactive waste is a sobering subject, but equally sobering statistics exist for other forms of toxic waste. According to the United Nations Environment Program, more than 440 million tons of hazardous waste are produced annually worldwide by oil refineries, chemical plants, hospitals, photo processing centers, laboratories, farms, dry cleaners, automobile repair shops, and a host of other industries and services. Much of that waste is stored in the ground in sealed containers. But much more is carelessly discarded into landfills and waterways, or is inadequately stored, and it finds its way, untreated and still dangerous, into the food chain. The effects of these wastes on human health, animal life, and the environment far exceed those of nuclear waste, as was seen at the Love Canal neighborhood of Niagara Falls, New York, where 21,000 tons of toxic waste caused such human damage that a portion of the city was demolished and abandoned.

This is cold comfort. We are steadily poisoning our planet with toxic waste dumped into the atmosphere, the ground, and the water supply. We demand electricity for a comfortable life, but we give little, if any, thought to the consequences of producing it.

CHAPTER 9

SUMMING UP

To readers who struggled through these 206 pages of complex, sometimes dense text referencing controversial and unresolved viewpoints, our admiration and gratitude. For those who, understandably, gave up en route, and especially those who headed directly to this chapter (perhaps a wise decision) we understand: life is short and linear quadratic equations long. To all, and to the casual reader, we owe a summary of what we discussed hcretofore. We also want to add some reflections and opinions of what are, admittedly, complicated and often controversial issues.

We wrote our book for several reasons. First and most important is that most people, even the best educated, know very little about radiation. Their prior exposure (excuse the pun), if any, may have been a high school or college course, now long forgotten. Second, most of us are unaccustomed to carefully weighing competing risks and benefits in making a choice between alternatives. This is not a shortcoming but rather because the human mind is imperfectly designed.

There are two fundamental problems. First, our brain is designed to process quickly large amounts of data, often held in cache, without conscious processes. This is a bit like what Watson (the IBM supercomputer) has to do when competing with human contestants on *Jeopardy!* Second, our brain is designed to weigh emotional before rational variables. (We do not imply emotions like love are always irrational, just usually.)

This first process, the rapid subconscious processing of large amounts of data, is referred to in psychology as *thin-slicing*. We arrive at conclusions quickly based mostly on impressions, which are often colored by prior experience. Thin-slicing is a consequence of evolution: not much time to evaluate a saber-tooth tiger's dinner plans when he appears outside your cave. Interestingly, these quick conclusions often prove as valid as more detailed, time-consuming analyses. Sometimes, however, they do not. When thin-slicing fails to reach the correct conclusion, results can be unfortunate or even disastrous. This is termed a Warren Harding–type error, referring to the twenty-ninth president of the United States, who gave an excellent immediate appearance but was less than effective in office. Consideration of complex data, like those related to radiation, sometimes displaces this quick conclusion process, but only rarely. Moreover, once we have reached a conclusion, we have difficulty changing our opinion. We selectively seek additional supporting data of whatever quality; discordances are often avoided or conveniently downgraded. If you doubt this process, consider the current debate over evolution.

Our second limitation is weighing emotional before rational considerations. Humans subconsciously weigh risks and benefits between alternatives, such as the use of fossil fuels versus nuclear energy to generate electricity. If choosing one option over another seems to offer little or no benefit, the risk of the latter is automatically perceived to be bigger. Subconsciously we think: nuclear energy has large risks, whereas fossil fuels have few, if any. Part of this subconscious appraisal has to do with the notion of what is "natural." Fossil fuels seem natural, organic (they are), much like drinking wheatgrass juice or eating crunchy granola, while nuclear energy seems unnatural, man-made. (It is! Although the fuel is not.)

Another aspect of our appraisal of benefit versus risk has to do with whether the risk is voluntary. Most people feel using nuclear energy is an involuntary risk imposed on them by government and/or industry. Curiously, they lack similar feeling about using fossil fuels. This makes the risk of nuclear energy seem scarier. Considerable data indicate societies are willing to accept 100- to 1000-fold greater risk if the risk is voluntary versus involuntary—riding a motorcycle, for instance, versus riding a bus. Finally, risk perception also depends on trust. A risk imposed by a distrusted source feels more threatening than a risk imposed by a trusted source. Again, the nuclear establishment (government, industry, etc.) is considered less trustworthy than the coal or oil industries and regulatory bodies. Why this is so is unclear; consider the *Deepwater Horizon* accident in the Gulf of Mexico or the *Exxon Valdez* oil spill in Alaska.

Still, shouldn't our powers of reason be able to overcome these instinctive impediments to clear thinking? No. In the complex interplay of slower, conscious reason and quicker, subconscious emotion and instinct, the basic architecture of the brain ensures we give more weight to emotion. The amygdala, where emotion is sited, receives incoming stimuli before the cortex, where we think things out. As the neuroscientist Joseph E. LeDoux of New York University puts it in *The Emotional Brain,* "The wiring of the brain . . . is such that connections from the emotional systems to the cognitive systems are stronger than connections from the cognitive systems to the emotional systems."

You may rightly ask: what has this to do with our book? A lot. When asked questions such as: Is exposure to radiation good? Is nuclear power safe? Should we abolish nuclear weapons?, most people resort to thin-slicing. Only later, and probably reluctantly, might they consider data supporting or contravening their first impression. We hope to help with this reprocessing of opinion by providing an unbiased analysis of data, albeit simplified, to help the reader through this more complex analytical process. We accept the dangers of simplification and have duly prepared ourselves for criticism from our scientific colleagues and persons who have far greater expertise in these complex subjects than we do. We also expect criticism from people with strong preformed opinions on issues such as nuclear energy and nuclear wastes. If we are attacked equally from both sides, we will judge ourselves to have hit the mark.

Because radiation touches every aspect of our lives—it is, in fact, responsible for our lives—it is essential to know what radiation is, how it works and what it can and cannot do.

Our belief in democracy is another reason we wrote this book. For democracy to function we need to be intelligently informed about issues we are asked to decide on. Quite often we vote for or against people or policies that touch on radiation-related issues. Examples include energy policy (Arctic drilling and licensing nuclear power facilities) and foreign policy, such as what to do about nuclear weapons development in North Korea and Iran. There are many more examples.

So, a few concluding thoughts and opinions. First, most peoples' fear of radiation is disproportionate to the real risk. We are all normally exposed to very different doses of background radiation depending on where we live and work.

Within this range, however, few data suggest adverse health effects. This means most of us should not worry about radiation from microwave ovens, TVs, computer screens, or cell phones. We should also not worry, as individuals, about backscatter X-ray screening at airports and other very low level radiation exposures. You will receive much more radiation during your flight than in the airport. (This does not mean that society should disregard the potential risks of exposing millions of people to even very small doses of radiation.)

What you should pay attention to is something most of us ignore: exposure to radiation from medical procedures. One-half of our annual average radiation exposure comes from medical procedures of which one-third or one-half may be unnecessary. Ask your physician why he/she is recommending a test and about the balance of risk and benefit and the dose of radiation you will receive. (This caution applies to all medical tests and procedures, not just radiation.) And if you live in certain areas of the United States and Europe you should also think about your possible exposure to radiation from radon gas. Measure levels in your home, if appropriate. Needless to say, if you are worried about cancer, don't smoke. Smoking not only exposes you (and your family members and friends) to substantial radiation, it works with radiation to cause cancer.

What about nuclear energy? We are neither proponents nor opponents. What we urge is a careful and critical evaluation of risks and benefits of each energy generating source, such as nuclear, fossil fuels, hydroelectric, solar, and others. We discuss, albeit briefly, the up- and downside of each. And remember, there are "nuclear risks" associated with non-nuclear energy sources, such as dependence on foreign oil, which could lead us to a military confrontation where tactical or strategic nuclear weapons are used. For example, is it wise for Japan, an island nation with few natural energy resources, to depend on external supplies given its long-term tense relations with China? Might this not trigger a political confrontation more dangerous than the risk of another Fukushima Daiichi accident? We, of course, don't know, but this type of complex scenario warrants careful consideration.

But if we promote the diffusion of peaceful nuclear technology, there will likely be an increase in the spread of nuclear weapons technology and capability. This confounding is unavoidable but probably unstoppable. Developing countries need cheap, efficient energy sources. India, with one in every five people on earth and an unreliable electrical grid is but one example. Then there is the issue of nuclear wastes. We believe this is a soluble problem scientifically but one that is stuck in a political morass. Definitive leadership is needed to progress. Politicians today need to take the "heat" to benefit future generations by tackling this issue. The payoff for an intelligent energy policy is fifty years down the road. Most of us will not reap the benefit of the tough decisions we need to make today. The question is whether politicians and energy company CEOs have the discipline and fortitude. The track record is not encouraging.

Nuclear accidents are certainly on many people's minds, especially in light of Chernobyl and Fukushima Daiichi. These accidents can have dreadful consequences. Fortunately, they are rare. But in considering these tragic events, we should also recall the relentless loss of life from practices such as the mining, transport, and burning of coal, not the least of which are air pollution and global warming. Increased radiation exposure from thinning of the ozone layer of the atmosphere could result in far more cancers than from a large nuclear accident. There is no easy solution to our insatiable energy appetite, and the best answer will likely lie in some combination of energy sources.

There is little to say in favor of nuclear weapons. We

can debate whether they were an effective deterrent of war between the United States and the Soviet Union, but there is no certain answer. Some people believe nuclear weapons should be completely eliminated. This concept has the advantage of disarming the argument that our friends but not our enemies should have them. A goal noble but naïve. What is more likely to happen is that our enemies will have nuclear weapons but not us. One has only to consider recent events in North Korea and Iran. Is it a higher moral ground to bomb the living daylights out of Pyongyang or Tehran with conventional weapons to prevent them from acquiring nuclear weapons? But we certainly don't need sixty-five thousand nuclear warheads when just a few will do. Also, neither of us would like to be the American president or Israeli prime minister on duty when a country like Iran develops a nuclear weapon. Definitive action is needed now. What that action should be is complex. Some have argued an attack will accelerate rather than retard rogue states from developing nuclear weapons. We disagree. Don't expect rational behavior from madmen, desperate people, or dictators.

There remain many important radiation-related issues we covered superficially or not at all. This was, in part, deliberate: most of us can take only so much science in one sitting. We wanted to give our readers an overview of many related topics rather than a drill-down on one or two. The interested reader will find more detailed information on our website and in the many in-depth books that focus on the topics of our chapters.

Again, thank you for staying with us until the end. The section that follows answers some frequently asked questions. More questions and answers are available on our website: www.radiationbook.com.

QUESTIONS AND ANSWERS

If radiation causes cancer, do I have to worry about every bit of radiation I am exposed to, including radiation from microwave ovens and cell phones?

It is highly unlikely every kind of radiation causes cancer. The electromagnetic waves used in microwave ovens and cell phones have not conclusively been shown to cause harm in humans.

Cell phones emit low-energy electromagnetic waves, which are non-ionizing and which, unlike some high-energy forms of UV radiation, are very unlikely to be able to cause chemical changes in living cells. The only known effect of such radiation is heating, much like microwaves in an oven heat water in foods. Because this is the only proved effect of these electromagnetic waves there is no known biological mechanism by which cell phone radiation could cause cancer in humans. Data regarding an increase in brain or other related cancers (meningiomas, gliomas, and cancers of the ear and salivary glands) in cell phone users are unconvincing, and there is no well-documented increase in any form of cancer despite the widespread increase in cell phone use over the past decade. The U.S. Food and Drug Administration, the Centers for Disease Control and Prevention, and the Federal Trade Commission do not classify cell phone use as dangerous to humans. In contrast, the International Agency for Research on Cancer and the American Cancer Society classify cell phone use as "possibly carcinogenic to humans." Everyone agrees more research on the safety of cell phones is warranted. If you are concerned about the danger of cell phones you may want to consider using a Bluetooth or similar device so that the antenna (the unit emitting electromagnetic waves) is farther from you than it is when holding a cell phone to your ear. Additional information from the U.S. National Cancer Institute is available at: http://www.cancer.gov/cancertopics/factsheet/Risk/cellphones.

If people in some places get much more radiation from natural background sources than in other places, do I need to move?

No. People in Denver receive more background radiation than people in New York. Denver residents get more radiation from the Sun (one mile high versus sea level) and from the Earth (the Rocky Mountains contain much more uranium, thorium, and radon than does the land in New York). But there are no data suggesting that Denver residents have a higher risk of cancers or other health problems than New Yorkers.

If nuclear energy has so many problems, including accidents and radioactive wastes, why don't we simply get rid of it?

Many countries get a substantial part of their electricity from nuclear power (20 percent in the United States, 30 percent in Japan, 80 percent in France). It would require an enormous amount of planning to switch to other energy sources, such as coal, without causing major economic and social disruption. Second, other energy sources have their own problems, such as pollution of the environment, global warming with coal and oil, and foreign energy dependence with oil. Worldwide, governments spend about half a trillion dollars annually on fossil fuel subsidies. (The United States has about 4 percent of the world's population and uses about 25 percent of the world's oil, coal, and natural gas.)

Can nuclear accidents like Chernobyl and Fukushima be prevented?

Yes, if we know and understand their vulncrabilities in advance, but we did not see these two coming. We have the capacity to design and control any man-made hazard once we truly understand it and accept its likelihood of occurrence, so likelihood can be substantially reduced. More advanced reactor designs will rely less on human intervention for safety. Reactors with this advanced design are being built in the United States and many other countries. We can't totally prevent airplane accidents, yet we don't abandon flying—we try to determine why accidents happen and take preventive steps. The same applies to nuclear power plants. We take serious precautions but have to acknowledge accident risk cannot be reduced to zero.

How much does exposure to radiation increase my cancer risk?

That depends on how much radiation you are exposed to. It is important to remember that all of us, alas, have a high cancer risk, whether or not we are exposed to extra radiation. A 50-year-old U.S. male currently without cancer has a more than 45 percent likelihood of developing one or more cancers in his remaining lifetime; this figure is about 38 percent for a woman. Exposure to radiation will increase this risk, but unless the dose is extremely high—a dose unlikely to be received by a member of the public—this increased cancer risk will be very small. The increased risk of cancer death from receiving an extra millisievert of radiation is similar to the actuarial risk of dying in one hour of canoeing or driving 300 miles.

Can we solve the problem of safely dealing with radioactive wastes?

Yes, but our approach needs to be driven by science, not by politics. There are several possible storage sites, though none is perfect, and a resolution of this challenge will likely require a combination of strategies. The problems we face in dealing with radioactive wastes are quite similar to those of other energy sources. The upside of radioactive waste is that the volume is relatively small. The downsides are the long life of some wastes and risk of diversion to nonpeaceful purposes.

What is the difference between external and internal radiation?

External radiation refers to a situation where a radiative source is external to the body. The source can be natural, as with cosmic or terrestrial radiations, or man-made, such as an X-ray machine, a CT scanner, or a radiation therapy device. Particles and/or electromagnetic waves from the radiation source passing into or through the body can cause changes in some tissues and organs. However, except for radiation from neutrons (a rare event), exposure to external radiation does not make us radioactive.

Internal radiation differs in that radioactive materials enter or are introduced into the body. Examples include breathing or eating air, food, or water containing one or more radionuclides. Sometimes radionuclides are injected into or placed inside us. For example, physicians sometimes inject patients with radioactive glucose to do a PET scan or with iodine-131 to do a thyroid scan. Some people with cancer have radioactive pellets of cesium-137 temporarily placed inside their body. This is a common therapy for breast and prostate cancers. When radioactive materials enter us, we become "radioactive." How long these radioactive substances remain inside us depends on many variables, including their physical half-life (their rate of radioactive decay) and their biological half-life (how long they remain in the body). People with radioactivity from medical procedures are not a risk to others except under special circumstances. It is important to recall we are all normally "radioactive" from naturally occurring radionuclides within us, such as potassium-40 and carbon-14. Ingesting radioactive food or water simply makes us more radioactive. Bon appétit.

Why is there so much controversy over irradiated foods?

Most foods contain infectious agents such as bacteria and fungi. Removing all of them by conventional sterilization techniques is often impossible, because it would destroy the food (or the food's nutritional value), or because it is economically unfeasible, or both. High doses of radiation are typically used to sterilize some foods. This process saves millions of lives worldwide; many people died from bacterial contamination of food before radiation sterilization techniques were developed. However, some people think that food sterilized by radiation is radioactive. This is wrong. Foods are typically sterilized with high doses of X-rays or gamma rays. Like any form of external radiation, these rays pass through the food but do not make it radioactive. Other people worry that irradiated food contains toxic substances, called free radicals, that may adversely affect our health. Although this is a theoretical possibility, all foods contain naturally occurring cancer-causing substances like free radicals (as well as substances that prevent cancer). The risk of an adverse outcome from eating irradiated food is far less than the risk of death from eating bacteria-contaminated food. Several recent outbreaks of toxin poisoning from *E. coli* contamination of meat and *Salmonella* contamination of eggs are good examples. People in the European Union seem especially (and inappropriately) concerned about food irradiation. Irradiated foods in the EU are required to have a special label.

If radiation is dangerous, should I refuse to have an X-ray study, a CT scan, or a mammogram? How about dental X-rays?

Because every exposure to ionizing radiations confers an increased cancer risk, however small, there should always be a good reason why a radiological study is done. If results of a study are likely to be useful, the study may be justified. However, if results to a patient have no practical utility for diagnosis and/or therapy, there is no justification for performing it. When a physician or dentist suggests a radiological study, you should ask the purpose, how the result will benefit you, and how much radiation you will receive. You can then make an informed decision as to whether the study is an appropriate balance of benefit and risk. For example, should you have a regular colonoscopy (which can be painful) to screen for colon cancer, or should you have a "virtual" colonoscopy, which is painless but exposes you to about 10 mSv of radiation—nearly two times your annual dose and hundreds of times more than you receive each year from the nuclear fuel cycle? You could spend decades continually going through a backscatter scanner at an airport before you received a dose equivalent to a virtual colonoscopy. Moreover, done as prescribed, you would need five of them in your lifetime, for a total exposure of 50 mSv—the maximum exposure allowed a worker in a nuclear power facility. Your physician or dentist should be able help you with this calculus. If not, find another doctor.

Is the type of radiation used to screen airline passengers dangerous?

Airport security personnel commonly use low-dose backscatter radiation to screen passengers for potentially dangerous objects. The dose from any one screening is 0.001 mSv, extremely small. A typical dental X-ray can deliver a dose of 0.01 mSv, which means it would take 100 backscatter screenings to equal one dental X-ray. All this accounts for a very small part of the radiation we receive every minute from natural background sources. Thus, being screened at an airport by this technique is probably safe. For those who believe in the linear, no-threshold radiation-dose hypothesis, when we multiply this small dose by the millions of people screened every day at airports worldwide, we will develop a number suggesting that some cancers will be caused by screening. But there is controversy as to whether this approach to estimating cancer risk is appropriate. Radiation protection professionals in the Health Physics Society and nuclear advocates from the American Nuclear Society argue that it is inappropriate to take a very small dose and multiply it by huge numbers of people. Others argue every radiation exposure can potentially cause cancer and that such a calculation is appropriate. The correct answer is unknown and probably unknowable. Certainly, the dose of radiation you get from backscatter screening is much lower than the radiation dose you will receive while flying any distance.

Do I need to worry that computer screens, TVs, and LED watches will increase my radiation dose and cause cancer?

No. Although small radiation doses are associated with these devices, the risk to any person is incredibly small (if any), no matter how much time you or your child spends playing computer games or watching TV.

Is nuclear energy our greatest nuclear threat?

We need to use nuclear energy carefully. However, there are probably greater nuclear threats to our lives—nuclear weapons, for example. Reliance on foreign sources of oil increases political instability, which increases the risk of war. In some instances, tactical or strategic nuclear weapons may come into play. The Middle East is a current source of political and economic instability and an important source of oil for the United States and many other countries. Asia is an even greater potential threat for instability. China and South Korea are building nuclear energy programs because of the risks inherent in foreign oil dependence, and they must establish and respect built-in safeguards. Nuclear terrorism remains a threat, as does the accumulation of nuclear weapons by unstable countries or even stable countries in unstable geopolitical regions, like North Korea and Iran.

What about nuclear terrorism?

Nuclear terrorism usually refers to the acquisition of nuclear materials or radioactive substances by persons not part of a defined nation-state. It can more broadly refer to rogue states that can use these materials to terrorize a neighbor, like Iran with Israel or North Korea with South Korea. Stolen or covertly developed nuclear materials can be used to make a bomb, to coat a conventional explosive device, or to contaminate a food or water supply. Nuclear weapons are, of course, dangerous, but a terrorist nonstate entity is most unlikely to be able to develop such a device. It is easier to imagine someone stealing or buying radioactive materials and converting a conventional explosive device. The actual radiological damage from such a device, termed an IND (improvised nuclear device) or RDD (radiological dispersal device), is likely to be small. But the psychological, political, and economic damage could be huge, largely because of public misunderstanding and exaggerated fears of the effects of radiation exposure, and because decontamination may be expensive and time consuming. Control of radioactive materials and public education are our best weapons against nuclear terrorism.

Can we evacuate large cities near nuclear power facilities quickly, or should those plants be shut down?

Politicians and opponents of nuclear energy often ask this question, based on the mistaken assumption that rapid evacuation of people living near a nuclear power facility is likely to be necessary or even desirable. Neither is true. A nuclear power reactor is not a nuclear weapon. It cannot explode like an atomic bomb, killing people by superfires and concussive force. When there is a radiation release, or the danger of a radiation release from a crippled nuclear power facility, we usually want to keep people *in* their home or office (referred to as "shelter in place") rather than evacuate them immediately, where they might be outdoors or in a car in a traffic jam when a radioactive cloud passes. Unless the level of radiation is so high that people will receive an unacceptable dose in a short interval, evacuation, if needed, should be delayed until a safe plan is developed and carefully implemented. Almost no scenario at a typical U.S. nuclear power facility would mandate immediate evacuation of a large urban population. Nevertheless, this argument was used to prevent commissioning the $3 billion Shoreham, New York, nuclear power facility, and it is an issue underlying some recent efforts to close the Indian Point, New York, facility. U.S. nuclear power stations are rarely sited very close to major population centers (although more are in western Europe). This is different, for example, from Ukraine, where there was a city (Pripyat) of 40,000 people within two miles of the Chernobyl nuclear reactor complex.

Isn't it dangerous when radioactive materials wind up in the ocean?

Yes and no. If depositing radioactive materials in the oceans can be prevented, it should be. However, sometimes, as with the Fukushima accident, this is impossible. Fortunately, ocean deposition of radionuclides rapidly dilutes them within a vast amount of water. This markedly reduces their risk to living organisms, like fish and plants. Also, some potentially dangerous radionuclides, like cesium-137, resemble potassium. Consequently, cesium-137 must compete with billions and billions and billions of times more naturally occurring potassium to enter into a living creature. Also, it is relatively rapidly excreted even after it is absorbed. Thus, except for extraordinary circumstances, ocean deposition of radioactive materials is less dangerous than it first appears. We should not forget that many nations intentionally dumped large amounts of radioactive materials (including scuttled nuclear submarines) in the oceans from 1946 to 1972. This practice has stopped. Also, atmospheric testing of nuclear weapons from 1945 to 1980 deposited substantial amounts of radioactivity in the oceans. Finally, let's not forget that the oceans normally contain radionuclides that enter them from various sources like earthquakes and runoff from rivers, which emerge from the ground carrying radioactive elements, including uranium, thorium, and radium. And they normally contain large amounts of naturally occurring potassium-40.

Should I take iodine tablets after a nuclear accident?

No, unless you are instructed to do so by your physician or a public health official. Nonradioactive iodine, usually potassium iodide (KI), is absorbed by the body and goes to the thyroid gland in varying amounts, depending on several factors, including how much iodine is in a person's normal diet. If the thyroid gland is filled with normal iodine, it is less likely to absorb radioactive iodine, like iodine-131, released by a nuclear accident. However, too much normal iodine can harm some people. Also, children can be poisoned if they accidentally take too much normal iodine. Finally, normal iodine is effective in blocking uptake of radioactive iodine only if it is taken *before* exposure to the radioactive form. Because of these complex considerations, people should take normal iodine tablets or syrup only when instructed to do so.

If radiation can cause cancer, why is it given to treat cancer?

Exposing normal cells to ionizing radiations has paradoxical effects. Low doses cause few changes, whereas increasing the dose increases the likelihood of damage to a cell's DNA. Damage to DNA (a mutation) can, under special circumstances, cause cancer. However, as the radiation dose increases further, a cell's DNA sustains so many mutations that it cannot survive and dies. From a cancer risk perspective, this is fine: a dead cell cannot cause cancer. (If it is a brain cell, this is not so good.) Cancer radiotherapy typically involves targeting extremely high doses of radiation to a specific body site, usually the site of the cancer. (There are some exceptions.) These high doses are intended to kill all cancer cells in the radiation field. Any normal cells in the field will also be killed. This strategy, in lung cancer and Hodgkin disease, can eradicate all cancer cells and cure people. Some previously normal cells may not be killed by radiation, and some cells outside the intended radiation field that receive scatter radiation may develop mutations resulting in a new cancer. For example, 5 to 10 percent of people cured of Hodgkin disease eventually develop leukemia or myelodysplastic syndrome. However, the benefits of using radiation to cure some cancers far exceed the risks of causing a new cancer.

Does exposure to man-made sources of radiation save or cost lives?

Modern life would be much less safe were it not for the widespread use of man-made sources of radiation, such as various detectors, exit signs, and industrial X-rays used to guarantee the structural integrity of aircraft and steel in buildings. Radiological studies are used to diagnose cancers and other diseases and to treat cancer. Overall, man-made sources of radiation save many more lives than they harm.

Why are people afraid of radiation?

A complex question. When radiation was first discovered, people were quite excited about its potential benefits. Many intentionally sought out places, like radon caves, where they could be exposed to radiation for supposed improved health. However, after the atomic bombs were dropped over Hiroshima and Nagasaki, things changed quickly. People started to see radiation as a danger rather than a benefit. This feeling was exacerbated during the Cold War nuclear arms race. Other negative influences include secrecy surrounding the development of nuclear submarines and warships and the belated recognition of the adverse health effects of radiation from atmospheric testing of nuclear weapons. Another factor in humans' fear of radiation is that we cannot detect it. Humans have evolved techniques to deal with fire, earthquakes, floods, and other natural disasters, but radiation is undetectable to our senses. A person can be exposed to a lethal dose of radiation and not even know it until days or weeks later. This concept frightens most people: better the devil we know than the one we don't. Most people know that radiation can, under special circumstances, cause cancers, birth defects, and genetic abnormalities, but they do not appreciate that other energy sources such as coal, gas, and oil also pose these dangers. Very few people realize, for example, that, per megawatt of electricity generated, coal-fired power facilities expose the public to three times more radiation than do nuclear power facilities. Also, most people do not realize how many sources of radiation they are exposed to in their daily lives. Finally, people tend to view nuclear power as scarier than its fossil fuel counterparts, in much the same way as many people fear dying in a plane crash more than in a car crash. Far fewer people die in plane crashes each year than in car crashes, and in exponentially fewer incidents, but the average number of deaths from a single plane crash is so much higher than in an average car crash that flying seems the riskier option. In essence, people are more afraid of outliers than of averages. In only one instance have many deaths (the 29 in the Chernobyl facility) been attributable to nuclear power, but that single outlier seems scarier than the vastly more numerous but smaller incidents involving fossil fuels, which have many inherent dangers.

How dangerous is radon gas in my home?

The greater the concentration of radon in the air and water at your home, the greater the potential danger. Radon accounts for about half of the average American's background radiation exposure. However, it is not uniformly distributed. Some areas in the United States, such as the Colorado plateau, have high levels because of the composition of radionuclides in the soil, whereas other areas have lower levels. People living in areas of high radon concentration should consider home radon testing. For information on how to test your home for radon, go to the Environmental Protection Agency website, http://www.epa.gov/radon/pubs/citguide.html. It is worth remembering that radon exposure may account for an important proportion of lung cancers and is especially dangerous to smokers.

Do astronauts and cosmonauts need to worry about radiation?

Yes. A lot. Space travel involves many complex risks, one of which is exposure to the same ionizing radiations we are exposed to on Earth, mostly X-rays and gamma rays from photons that make it through our atmosphere (radon is a different issue). However, similar to people living in Denver and airline flight crews who receive more background radiation than most of us because they are closer to the Sun, astronauts, such as those living on the International Space Station (ISS), receive a higher radiation dose. For example, a six months' residence in the ISS results in a dose of about 120 mSv, or about one-half of the dose of the average atomic bomb survivor. This is about sixty times more than they would receive if they remained on Earth. Cosmonauts on long space missions, especially inter-planetary travel, will receive much higher doses. For example, persons going to and (hopefully) returning from a trip to Mars might receive a dose of 1,000 mSv or five times as much as the average atomic bomb survivor.

There are three sources of radiation cosmonauts will be exposed to: (1) solar particle events (coming from our Sun), which occur episodically, vary with the solar cycle (solar flares and coronal mass ejections), and can be only partially predicted; (2) galactic cosmic events (coming from the Galaxy exclusive of our Sun), which are high-energy charged atoms stripped of their electrons, traveling near the speed of light and able to penetrate almost anything; and (3) radiation belts trapped by the Earth's magnetic field. The Van Allen radiation belt is an example of these belts of energetic charged particles.

Moreover, although scientists have a reasonably good idea about the potential health consequences of some of the types of radiation cosmonauts will be exposed to, such as gamma rays and protons, very little is known about the potential health effects of charged ions heavier than hydrogen (which makes up protons). Two major types of health events need to be considered: (1) short-term events, such as damage to the bone marrow, gastro-intestinal tract, eyes (cataracts), skin, heart, and central nervous system; and (2) cancers. Potential alterations in behavior are also of concern.

NASA and the space agencies are involved in research in this area, including optimal spacecraft design for radiation shielding, better knowledge of the health effects of exposure to charged particle radiations, and selection of crew members with the lowest risk of cancer development. For example, the recent Mars mission capsule and Curiosity rover contain devices to measure radiation doses astronauts would receive traveling to and operating on Mars.

Does everyone exposed to the same dose of radiation have the same risk of developing cancer?

No. We discussed several important variables such as age and gender. For example, young people are more likely to develop thyroid cancer from exposure to iodine-131 than are adults. Also, women are more likely to develop it after radiation exposure than are men. Other factors, however, also affect risk. People with certain genetic disorders, especially disorders in which the ability to repair damage to DNA is compromised, are especially susceptible to radiation-induced cancers. These include children with ataxia telangiectasia (which affects the brain and other parts of the body), Fanconi anemia (an inherited blood disorder), and xeroderma pigmentosum (the inability of the body to repair damage done by UV light). Normal-appearing parents of these children who have one copy of the abnormal gene (usually the children have two copies), may also have an increased susceptibility to radiation-induced cancers. Other people with no family history of a genetic disease may have variants of genes or mutations in genes (or both) that increase sensitivity to radiation. For example, female radiology technicians with the BRCA2 gene are at increased cancer risk compared to female radiology technicians who lack the BCRA2 gene. (BCRA2 is an inherited mutation associated with an increased risk of breast and ovarian cancer.) And women with the BCRA2 gene mutation, who are especially sensitive to radiation-induced breast cancer, may further increase their breast cancer risk by having frequent mammograms. (This does not mean they should forego mammograms, simply that having them adds an increased risk, compared to women without the BCRA2 gene mutation.) There are many other associations, such as the increased risk of lung cancer in atomic bomb survivors with a certain form of a gene that encodes the epidermal growth factor receptor, a protein that regulates some types of cells in the lung.

The bottom line is that different people have different susceptibilities to radiation-induced cancers. At present, it is difficult to know what a person's relative sensitivity is unless there is a family history of a distinct genetic abnormality. Advances in gene sequencing, however, are likely to alter this picture over the next decade. But we need to recall that there is also an element of randomness as to who develops a radiation-induced cancer. So both variables operate. Regardless of a person's underlying genetic susceptibility, there is a clear relationship between radiation dose and cancer risk: the greater the dose, the greater the risk. This issue of genetic susceptibility to cancer development from radiation is important when we consider, for example, who to send into space for inter-planetary travel where high-dose radiation may be unavoidable.

For more information and links to articles, studies, and visual aids related to this book, please visit our website: www.radiationbook.com.

ACKNOWLEDGMENTS

Our book deals with many complex and controversial issues. Some lack a clear answer. We sought information, comments, and critiques from expert colleagues. All gave sound advice, much of which we accepted. For friendly suggestions, guidance, and forbearance, we thank Leslie Botnick of Vantage Oncology; Neil Buist, M.D., FRCPE, Professor Emeritus, Pediatrics, Oregon Health and Science University; Howard W. Dickson, CHP, President, Dickson Consulting, L.L.C.; Robert J. Emery, Ph.D., Professor of Occupational Health, The University of Texas School of Public Health; F. Owen Hoffman, President and Director SENES Oak Ridge, Inc. Center for Risk Analysis, Distinguished Emeritus Member of the National Council on Radiation Protection and Measurements, and Consultant to the United Nations Scientific Committee on the Effects of Atomic Radiation; Waltraud Holzer, the former Executive Secretary at the United Nations Scientific Committee on the Effects of Atomic Radiation in Vienna; Bennett Ramberg, Ph.D, author of *Nuclear Power*

Plants as Weapons for the Enemy: An Unrecognized Military Peril; Eric R. Scerri, Ph.D., Department of Chemistry and Biochemistry, UCLA, author of *The Periodic Table, Its Story and Its Significance*; Arnold Sherwood, Ph.D.; Masao Tomonaga, M.D.; F. Ward Whicker, Ph.D., Professor Emeritus, Environmental and Radiological Health Sciences, Colorado State University; and Susan C. Winter, MD, FAAP, FFACMG, Medical Director, Medical Genetics/Metabolism, Children's Hospital Central California, Clinical Professor of Pediatrics, University of California/San Francisco. Their contributions improved this book greatly, and credit for what is correct is theirs. Any errors are ours alone.

Our thanks as well to Walter Isaacson; Diana K. Buchwald, General Editor and Director of the Einstein Papers Project at Caltech; Barbara Wolff of the Albert Einstein Archives, Hebrew University of Jerusalem; and Sarah L. Malcolm of the Franklin D. Roosevelt Presidential Library, for their assistance with permission to use the Albert Einstein letter to Franklin Roosevelt; and to Marc Bouquet, Ph.D., Centre d'Enseignement et de Recherche en Environnement Atmosphérique (CEREA), Paris, France, for permission to use his lab's illustrations on the dispersal of cesium-137 from Fukushima and Chernobyl.

At Knopf, our thanks to Jonathan Segal for his support for this book and for his always deft editing; to his assistant Joey McGarvey for her steady and cheerful help; to Victoria Pearson for her grace and humor in navigating the book through production on a schedule we made far too tight; to Amy Ryan for her diligent and thoughtful copy editing; to Janet Biehl for a multitude of last-minute heroics; and to Michelle Somers and Brittany Morrongiello for quick and inventive thinking.

RPG and EL

Many physicians and scientists collaborated with me over these years. It is impossible to name them all, but some deserve special mention. I apologize to anyone I missed; you know who you are. My colleagues at the University of California, Los Angeles, supported and contributed greatly to several emergency responses, including Professors Richard A. Champlin, now of the University of Texas, Paul I. Terasaki, Drew J. Winston, Winston G. Ho, Emmanuel Maidenberg, and the late Professors M. Ray Mickey and David W. Golde. Also, my mentor and friend Professor Martin J. Cline and the late Professor Alexander Friedenstein, who suffered greatly under the Soviet system. Professor Yair Reisner of the Weizmann Institute of Science, a lifelong colleague and friend, who braved the Cold War to help the Chernobyl victims. The late Dr. Armand Hammer and Richard Jacobs helped me immensely with resources. Professor Alexander Baranov of the Burnasyan Federal Medical Biophysical Center (FMBC) has been my colleague, teacher, and friend for more than twenty-five years. He and I worked with Professor Angelina Guskova and Academicians Andrei Vorobiev and Lenoid Illyn of the FMBC and Russian Research Center for Hematology, the late Georgi Selidovkin of the FMBC, and many other talented Russian and Ukrainian physicians and scientists. Academician Yergeniy Chazov allowed me to raid his institute for equipment or supplies on several occasions, and Dr. Vicktor Voskresenskiy, my KGB "minder," always helped me skirt the "system" when needed, despite his official role. President Mikhail Gorbachev, along with the late Ambassador Anatoly Dobrynin, were key in inviting me to the Soviet Union and encouraging my efforts. Foreign Minister Nikolai Ryzkov help with my efforts in Armenia after the earthquake in 1988. The late Professor E. Donnall Thomas of the University of Washington and his Seattle colleagues got me interested in bone marrow transplantation more than forty years ago. I was also helped early on by the late Professors Mortimer M. Bortin of the Medical College of Wisconsin, George W. Santos of Johns Hopkins University, and Georges Mathé of Institut Gustave Roussy.

In Brazil, I was privileged to work with Dr. Daniel Tabak, then of the National Cancer Institute, and with Admiral Amihai Burla of the Brazilian Navy who "liked" me so much he tried to prevent my exit by confiscating my passport on several occasions. My colleagues at the Naval Hospital Marcílio Dias in Rio and the Brazilian Nuclear Energy Agency were very important collaborators. Professor Roland Mertlesmann of Freiburg University also helped. Waltraud Holzer helped with contact to the International Atomic Agency for many years.

In Japan, I am indebted to Professors Kazuhiko Maekawa, Shigeru Chiba, and Shigetaka Asano of Tokyo University; Hakeumi Oh, Hideke Kodo, Hideo Mugishima of Nihon University; and Masao Tomonaga of Nagasaki University, who kindly shared his recollections of the atomic bombings and has dedicated his life to study the aftereffects. Professor Mine Harada of Kyushu University has been a lifelong colleague and friend and has always supported these complex projects. Ms. Noriko Shirasu helped with liaison to the Prime Minister's office and Diet. Celgene Corporation supported my humanitarian efforts in Japan led by Mr. Joseph Melillo and Drs. Jerry Zeldis and Jay Backstrom.

Most of all, I am indebted to my family, who supported me through these adventures. Tal, Shir, and Elan took their father's advice and braved Kiev soon after Chernobyl to help calm the population. My wife, Laura Jane, read many iterations of this book and kept Eric and me clothed and fed through endless sessions.

Lastly, may I recognize Andrei Tarmozian, a Chernobyl hero, who kept in touch for many years and who, despite his injuries, traveled around the world to discuss his experiences with fellow firefighters. He lived to see his grandchild and only recently died of causes unrelated to radiation. I shall miss him.

RPG

Jon Segal has been my editor for more than thirty years, through six books. He improves every manuscript he touches with his care for the writer and his attention to each detail. It is a given in publishing that he is an extraordinary editor. To my good fortune, he is also an extraordinary friend.

My gratitude to Laura Gale for her comments on the manuscript, as well as for her care, feeding, and friendship, while Bob and I worked in Los Angeles, New York City, Big Sky, Montana, and London. My gratitude as well to Danielle Flam for a careful and thoughtful reading. Great thanks to E. C. McCarthy, a wonderful writer who helped arrange a vast amount of material and who often came up with just the right word.

David Wolf and William Tyrer suffer through many drafts of every book I write and never say no to the next. Their decades of friendship are a steady boon, their camaraderie a joy. My sons Simon and John each made contributions with their reading, questions, and suggestions. One night over dinner, Simon offered an insightful theory of why people fear radiation. I asked him to write it out, and it stands largely intact in these pages. My wife, Karen Sulzberger, is an equally thoughtful reader and questioner, a fount of encouragement and support, and best of all, a glorious partner in love and in life.

EL

NOTES

INTRODUCTION. "THE CESIUM BOMB"

3 *In 1985 two radiologists in Goiânia:* For an account of the Goiânia accident, see International Atomic Energy Agency, *The Radiological Accident in Goiânia* (Vienna: IAEA, 1988), http://www-pub.iaea.org/mtcd/publications/pdf/pub815_web.pdf. Wikipedia has an interesting summary as well: http://en.wikipedia.org/wiki/Goiania. It includes details such as the guard going to a movie, and others translated from local Portugese-language newspapers. The IAEA report describes the thorough search for radioactive contamination of the city that was carried out. The home of one of the scrap dealers was demolished, and the materials and topsoil were buried in a secure place about 13 miles outside town, along with soil from other highly contaminated areas. In all, 85 homes were found to have some level of contamination; 41 were evacuated. All the houses were cleaned and decontaminated with a combination of vacuums with high efficiency filters, jet

spray washing, and chemical decontaminants. The 45 public places with traces of cesium dust were also decontaminated in the same manner.

9 *Because the victims*: For details of medical treatment of radiation victims, see Anna Butturini, Robert Peter Gale, et al., "Use of Recombinant Granulocyte-Macrophage Colony Stimulating Factor in the Brazil Radiation Accident," *The Lancet,* August 27, 1988.

12 *To have a sense of just how small*: For a fascinating video on atomic size, see Jon Bergmann, *Just How Small Is an Atom?* April 2012, animation by Cognitive Media, http://www.ted.com/talks/. The comparisons to a blueberry and compressing billions of cars are from this. Also see www.ted.com/talks/just_how_small_is_an_atom.html. For a sense of the size of the universe, see Cary and Michael Huang, "The Scale of the Universe 2," 2012, http://htwins.net/scale2/.

CHAPTER 1. ASSESSING THE RISKS

23 *sacred Hindu scripture*: Jeremy Pearce, J. Robert Oppenheimer obituary, *The New York Times,* November 22, 2004.

23 *apart from laboratory work*: It is dangerous to say anything was a first in Nature. There actually was fission on Earth before the A-bomb was created. Approximately 1.7 billion years ago at what is now Oklo, Gabon, naturally caused chain reactions occurred over a period of perhaps a few hundred thousand years in a large uranium deposit. After heavy rains, groundwater accumulated and acted as a neutron moderator; when the water boiled away, the chain reaction stopped until the next deluge.

36 *between 11,000 and 270,000 developed extra thyroid cancers:* F. Owen Hoffman, David C. Kocher, and A. Iulian

Apostoaei, "Beyond Dose Assessment: Using Risk with Full Disclosure of Uncertainty in Public and Scientific Communication," *Health Physics,* November 2001, vol. 101, issue 5, pp. 591–600.

CHAPTER 2. RADIATION FROM DISCOVERY TO TODAY

39 *The nineteenth-century pioneers*: http://www.nuclearfiles.org/menu/timeline/ is an excellent source for the history of radiation and of nuclear science, and we have relied on them for dates and achievements of the scientists mentioned. For those who are Nobel laureates, http://www.nobelprize.org has detailed descriptions of their work, and we have drawn on this source for their biographies. The Leó Szilárd quotes are from nuclearfiles.org.

44 *Radium Girls*: Bill Kovarik, "The Radium Girls," http://www.radford.edu/~wkovarik/envhist/radium.html.

49 *150,000 to 240,000 people:* Details of atomic bomb survivors have been taken from the Radiation Effects Research Foundation, Life Span Study report titles, http://www.rerf.or.jp/library/archives_e/lsstitle.html; Evan B. Douple et al., "Long-term Radiation-Related Health Effects in a Unique Human Population: Lessons Learned from the Atomic Bomb Survivors of Hiroshima and Nagasaki," *Disaster Medicine and Public Health Preparedness* 5, suppl. 1 (March 2011): S122–133; and National Health Council, *Health Levels of Exposure to Low Levels of Ionizing Radiation* (*BEIR VII*) (Washington, D.C.: National Academies Press, 2006).

51 *Albert Einstein, Old Grove Rd:* Letter, Albert Einstein to Franklin D. Roosevelt, August 2, 1939; President's Sec-

retary's File (Safe Files), Sachs, Alexander Index; The Franklin D. Roosevelt Library and Museum.

54 *Leukemia was the first*: Details of illnesses after the A-bombs are from: Kotaro Ozasa, Yukio Shimizu, Akihiko Suyama, Fumiyoshi Kasagari, Midori Soda, Eric J. Grant, Ritsu Sakata, Hiromi Sugiyama, and Kazunori Kodama, "Studies of the Mortality of Atomic Bomb Survivors, Report 14, 1950–2003: An Overview of Cancer and Noncancer Diseases." *Radiation Research* 177, 239–243 (2012), Radiation Research Society.

57 *45 percent of males*: American Cancer Society, http://www.cancer.org/Cancer/CancerBasics/lifetime-probability-of-developing-or-dying-from-cancer. The figure for women is 38 percent.

CHAPTER 3. THE NATURE OF RADIATION

73 *The natural decay chain of uranium-238*: New York State Department of Health, "Info for Consumers: Uranium-238 Decay Chain," March 2000, http://www.health.ny.gov/environmental/radiological/radon/chain.htm.

84 *the number at 1 million:* Brian Meonch, "Chernobyl Cover-up: Study Shows More Than Million Deaths from Radiation," *Independent Australia*, April 21, 2011, http://www.independentaustralia.net/2011/life/health/chernobyl-cover-up-study-shows-more-than-a-million-deaths-from-radiation/.

84 *The World Health Organization reported*: World Health Organization, "Health Effects of the Chernobyl Accident: An Overview," April 2006, http://www.who.int/ionizing_radiation/chernobyl/backgrounder/en/index.html.

85 *about 15 have died*: United Nations Scientific Committee

on the Effects of Atomic Radiation, "Sources and Effects of Ionizing Radiation," UNSCEAR 2008 Report to the General Assembly with Scientific Annexes, vol. 2, sections C, D, and E, New York: 2011, http://www.unscear.org/docs/reports/2008/11-80076_Report_2008_Annex_D.pdf.

85 *as many as 25,000 cancers*: World Health Organization, International Agency for Research on Cancer, "Briefing Document: The Cancer Burden from Chernobyl in Europe," April 2006, www.iarc.fr/en/media-centre/pr/2006/IARCBriefingChernobyl.pdf.

88 *170 million Americans*: Ann G. Moore, A. Iulian Apostoaei, Ph.D., Brian A. Thomas. M.S., F. Owen Hoffman, "Thyroid Cancer from Exposure to I-131 from the Nevada Test Site," Senes Oak Ridge, Inc., October 17, 2006, http://www.senes.com/Thyroid.Doses.Final.Report.pdf; and "Estimated Exposures and Thyroid Doses Received by the American People from Iodine-131 in Fallout Following Nevada Atmospheric Nuclear Bomb Tests (A report from the National Cancer Institute)" http://www.cancer.gov/i131/fallout/contents.html

94 *liquidators received an average*: United Nations Scientific Committee on the Effects of Atomic Radiation, "The Chernobyl Accident, UNSCEAR's assessment of the radiation effects," http://www.unscear.org/unscear/en/chernobyl.html

99 *GM-CSF proved useful*: Alexandr Baranov, Robert Peter Gale, et al., For details of the treatment of the Chernobyl victims, see "Bone Marrow Transplantation After the Chernobyl Nuclear Accident," *New England Journal of Medicine* (July 27, 1989): 205–12.

103 *"betrayed the nation's right"*: Hiroko Tabuchi, "Inquiry Declares Fukushima Crisis a Man-Made Disaster," *The New York Times,* July 5, 2012.

CHAPTER 4. RADIATION AND CANCER

105 *Now, however, it is believed*: Gina Kolata, "Bits of Mystery DNA, far from 'Junk' play Crucial Role." *The New York Times*, September 5, 2012, http://www.nytimes.com/2012/09/06/science/far-from-junk-dna-dark-matter-proves-crucial-to-health.html

107 *number of nonsmokers killed worldwide is about 600,000:* Miriam Falco, "Secondhand Smoke Kills 600,000 Worldwide Annually," *CNN Health,* November 26, 2010, Thechart.blogs.cnn.com/2010/11/26secondhand-smoke-kills600000-worldwide-annually/.

109 *Cigarette manufacturers have known:* Vincenzo Zaga, Charilaos Lygidakis, Kamal Chaouchi, and Enrico Gattavecchia, "Polonium and Lung Cancer," *Journal of Oncology* 2011, article ID 860103. Their paper is the source for risks associated with smoking.

110 *kill more than 1 million people annually:* Frederica P. Perera, "Molecular Clues to Preventing Tobacco-Related Lung Cancer," *Cancer Prevention,* no. 7 (Spring 2006); http://www.nypcancerprevention.com/issue/7/pro/feature/molecular-clues-to-preven.shtml.

113 *St. Louis physician Dr. Louise Reiss:* Dennis Hevesi, "Dr. Louise Reiss, Who Helped Ban Atomic Testing, Dies at 90," *The New York Times,* January 10, 2011, http://www.nytimes.com/2011/01/10/science/10reiss.html; and Michael D. Sorkin, "Louise Reiss: Headed Historic Baby Tooth Survey in St. Louis," *St. Louis Post-Dispatch,* January 7, 2011, http://www.stltoday.com/news/local/obituaries/article_bc5094d0-34b1-5c14-b4b0-fd8251cb7990html.

119 *200,000 cases of melanoma*: World Health Organization, http://www.who.int/mediacentre/factsheets/fs305/en

/index.html; *as much as 800 percent*: http://www.skincancer.org/skin-cancer-information/skin-cancer-facts.

119 *65,000 melanoma-related deaths worldwide*: World Health Organization, "Health Consequences of Excessive Solar UV Radiation," press release, July 25, 2006, http://www.who.int/mediacentre/news/notes/2006/np16/en/index.html.

119 *People in sunny climes:* David Schottenfeld and Joseph F. Fraumani, Jr., *Cancer Epidemiology and Prevention,* 2nd ed. (New York and Oxford: Oxford University Press, 1996), 356.

119 *Two studies*: Ibid., 357.

124 *2.8 million cases*: WHO, http://www.who.int/mediacentre/factsheets/fs305/en/index.html.

125 *psoriasis, a malady*: National Library of Medicine, U.S. National Institutes of Health, http://www.ncbi.nlm.nih.gov/pubmedhealth/PMH0001470/\\.

CHAPTER 5. GENETIC DISEASES, BIRTH DEFECTS, AND IRRADIATED FOOD

130 *3 percent of children*: Physicians Committee for Responsible Medicine, http://www.pcrm.org/search/?cid=2785

140 *48 million cases*: Centers for Disease Control and Prevention, http://www.cdc.gov/foodborneburden/ and United States Department of Agriculture, Agricultural Research Service, http://www.ars.usda.gov/main/site_main.htm?modecode=53-25-23-00.

140 *It is difficult to arrive at*: For more information about food irradiation, see Marler Clark, "About Foodborne Illness," n.d., *Foodborne Illness,* http://www.foodborneillness.com/; US Enviromental Protection Agency, "Food Irradiation," *RadTown USA,* August 14, 2012, http://www.epa

.gov/radtown/food-irradiation.html; Joe Schwarcz, "Good Old Days," Office for Science and Society blog, McGill University, November 4, 2011, http://oss.mcgill.ca/yasked/foodirradiation.pdf; and Radiation Information Network, "Food Irradiation," n.d., Idaho State University, http://www.physics.isu.edu/radinf/food.htm.

140 *2 billion cases*: Joint news release WHO/FAO, 11 October 2004, http://www.who.int/mediacentre/news/releases/2004/pr71/en/.

CHAPTER 6. RADIATION AND MEDICINE

149 *Allan Cormack (1924–1998)*: Details of the life and work of Allan Cormack and Godfrey Hounsfield can be found at Robert McG. Thomas Jr., obituary for Allan Cormack, *The New York Times,* May 1998; Ian Isherwood, obituary for Sir Godfrey Hounsfield, *Radiology* 234 (March 2005): 975–76, and Nobel Foundation, "The Nobel Prize in Physiology or Medicine 1979: Allan M. Cormack, Godfrey N. Hounsfield," Nobelprize.org., n.d.

150 *While taking a stroll in the countryside*: Details of life and work of Godfrey Hounsfield can be found at http://www.nobelprize.org/nobel_prizes/medicine/laureates/1979/hounsfield-autobio.html.

151 *A study published in* The Lancet: Mark S. Pearce, Jane A. Salotti, Mark P. Little, et al. "Radiation Exposure from CT Scans in Childhood and Subsequent Risk of Leukaemia and Brain Tumours: A Retrospective Cohort Study," *The Lancet,* August 4, 2012, 499–505.

CHAPTER 7. BOMBS

169 *65,000 nuclear warheads*: Brookings Institution, "Research: Defense and Security: Nuclear Weapons," 2012, http://www.brookings.edu/research/topics/nuclear-weapons. According to the Arms Control Organization, in addition to the 1,492 Russian and 1,731 U.S. operational warheads, Russia has 4,000 more that have not been deployed or are in storage, and the United States has about 2,700 in storage. The U.S. also has approximately 500 operational tactical weapons, about 200 of which are deployed in Europe. See Arms Control Association, "Nuclear Weapons: Who Has What at a Glance," press release, August 2012, http://www.armscontrol.org/factsheets/Nuclearweaponswhohaswhat.

171 *workers at a nuclear fuel–processing plant in Tokaimura:* S. Chiba, A. Saito, K. Takeuchi, R. P. Gale, et al., case report, "Transplantation for Accidental Acute High-dose Total Body Neutron–and γ-radiation Exposure," *Bone Marrow Transplantation* (2002) 29, 935–939.

173 *"instant incineration"*: Paul Vitello, obituary for Paul S. Boyer, *The New York Times,* April 2, 2012, http://www.nytimes.com/2012/04/02/us/paul-s-boyer-76-who-wrote-about-a-bomb-and-witches-dies.html.

175 *"It's hard to imagine any kind of dirty bomb":* http://www.cdi.org/terrorism/dirty-bomb.cfm and http://cees.tamiu.edu/covertheborder/TOOLS/NationalPlanningSen.pdf.

175 *Bob and Alexander Baranov wrote:* Robert Peter Gale and Alexander Baranov, "If the Unlikely Becomes Likely: Medical Response to Nuclear Accidents," *Bulletin of the Atomic Scientists* 67, no. 2 (2011): 10–18.

177 *"designed to withstand hurricanes, earthquakes"*: Carl Behrens and Mark Holt, "Nuclear Power Plants: Vulnera-

bility to Terrorist Attack," Congressional Research Service Report to Congress, February 4, 2005, http://www.fas.org/irp/crs/RS21131.pdf.

178 *In late 1983*: Sandra Blakeslee, "Nuclear Spill at Juarez Looms as One of Worst," *The New York Times*, May 1, 1984.

179 *In 1998:* Nuclear Free Local Authorities, "Radioactive Scrap Metal," n.d., http://www.nuclearpolicy.info/publications/scrapmetal.php; and International Atomic Energy Agency, "IAEA Conference on 'Safety of Radiation of Radiation Sources and Security of Radioactive Materials,' " September 14–18, 1998, Dijon, France, http://www-ns.iaea.org/meetings/rw-summaries/dijon-1998.asp.

179 *The reports showed how often radioactive material*: International Atomic Energy Agency, "IAEA Conference on 'Safety of Radiation Sources and Security of Radioactive Materials,' " September 14–18, 1998, Dijon, France, http://www-ns.iaea.org/meetings/rw-summaries/dijon-1998.asp.

CHAPTER 8. NUCLEAR POWER AND RADIOACTIVE WASTE

183 *In a 2009 article in* The Lancet: Anil Markandya and Paul Wilkinson, "Electricity Generation and Health," *The Lancet,* September 15, 2007, 979–81.

189 *A normally operating nuclear power plant*: Mara Hvistendahl, "Coal Ash Is More Radioactive Than Nuclear Waste," *Scientific American,* December 13, 2007, http://www.scientificamerican.com/article.cfm?id=coal-ash-is-more-radioactive-than-nuclear-waste; and Hans-Joachim Feuerborn, "Coal Ash Utilisation over the World and in Europe" (paper, Workshop on Environmental Health Aspects of Coal Ash

Utilization, November 23–24, 2005, Tel-Aviv, Israel), www.scribd.com/doc/90578989/abstract-Feuerborn.

190 *Grand Central Terminal*: http://www.pbs.org/wgbh/pages/frontline/shows/reaction/interact/facts.html.

190 *24 tons of waste*: Nuclear Energy Institute, "Nuclear Waste: Amounts and On-Site Storage, http://www.nei.org/resourcesandstats/nuclear_statistics/nuclearwasteamountsandonsitestorage.

191 *The burning of coal produces*: CBS News, "Coal Ash: 130 Million Tons of Waste," *60 Minutes,* August 15, 2010, http://www.cbsnews.com/2100-18560_162-5356202.html and Brian Merchant, "Nuclear Waste Piling Up Across US: 138 Million Pounds and Counting," Treehugger, November 17, 2010, http://www.treehugger.

191 *130 million tons of ash*: CBS News, *60 Minutes,* August 15, 2010, http://www.cbsnews.com/2100–18560_162–5356202.html.

191 *200 feet on a side*: The Nuclear Energy Institute estimates that the 67,500 metric tons of all used fuel assemblies from the past four decades would cover a football field to a depth of 21 feet. http://www.nei.org/resourcesandstats/nuclear_statistics/nuclearwasteamountsandonsitestorage/.

191 *nuclear reactors designed to operate at very high temperatures:* www.eoearth.org/article/Hydrogenproduction_from_nuclear_power.

194 *Nuclear fuel consists*: We are grateful to F. Ward Whicker, Ph.D., for his advice about this discussion of radioactive waste.

201 *more than $60 billion*: F. W. Whicker, T. G. Hinton, M. M. MacDonnell, J. E. Pinder III, and L. J. Habegger, "Avoiding Destructive Remediation at DOE Sites," *Science* 303 (March, 12, 2004): 1615–1616.

202 *it was "regarded primarily as garbage"*: U.S. Environmental

Protection Agency, National Service Center for Environmental Publications, "Fact Sheet on Ocean Dumping of Radioactive Waste Materials," November 20, 1980.

203 *14 zetabecquerels*: Radiation Information Network, "Radioactivity in Nature," Idaho State University, n.d., http://www.physics.isu.edu/radinf/natural.htm, based on figures from the National Academies of Science. Most of the oceans' radioactivity is from potassium-40, ten times the amount from rubidium-87, followed by small amounts of uranium, carbon-14, and tritium.

203 *4.5 billion tons of uranium*: Will Ferguson, "Record haul of uranium harvested from seawater," *New Scientist*, August 22, 2012, http://www.newscientist.com/article/dn22201-record-haul-of-uranium-harvested-from-seawater.html.

205 *440 million tons of hazardous waste*: Meena Palaniappan et al., "Cleaning the Waters: A Focus on Water Quality Solutions," report to the United Nations Enviroment Programme, March 2010, Nairobi, Kenya, www.unwater.org/downloads/Clearing_the_Waters(1).pdf.

CHAPTER 9. SUMMING UP

209 *"The wiring of the brain"*: David Ropeik, *The New York Times,* September 28, 2012, http://www.nytimes.com/2012/09/30/opinion/sunday/why-smart-brains-make-dumb-decisions-about-danger.html.

SELECTED BIBLIOGRAPHY

Adami, Hans-Olov, David Hunter, and Dimitrios Trichopoulos, eds. *Textbook of Cancer Epidemiology.* Oxford: Oxford University Press, 2002.

Chin, John L. and Allan Ota. "Disposal of Dredged Materials and Other Waste on the Continental Shelf and Slope." In Herman A. Karl, John L. Chin, Edward Ueber, Peter H. Stauffer, and James W. Hendley II, eds., *Beyond the Golden Gate: Oceanography, Geology, Biology, and Environmental Issues in the Gulf of Farallones.* Circular 1198. Menlo Park, Calif.: U.S. Geological Survey and U.S. Department of the Interior, 2006. http://pubs.usgs.gov/circ/c1198/chapters/193-206_Disposal.pdf.

Fraley, L., Jr., and F. W. Whicker. "Response of Shortgrass Plains Vegetation to Gamma Radiation—II. Short-Term Seasonal Irradiation." *Radiation Botany* 13, no. 6 (December 1973): 343–52.

Fraley L., and F. W. Whicker. "Response of a Native Shortgrass Plant Stand to Ionizing Radiation." In D. J. Nelson, ed., *Radionuclides in Ecosystems: Proceedings of the Third*

National Symposium on Radioecology 999–1006. Washington, D.C.: U.S. Atomic Energy Commission, 1971.

Institute of Medicine, Committee on Thyroid Screening Related to I-131 Exposure, Board on Health Care Services. *Exposure of the American People to Iodine-131 from Nevada Nuclear Bomb Tests.* Washington, D.C.: National Academies Press, 1999.

International Atomic Energy Agency. "Effects of Ionizing Radiation on Plants at Levels Implied by Current Radiation Protection Standards." Technical Report Series 332. 1992, Vienna.

International Atomic Energy Agency. *Environmental Impact of Radioactive Releases: Proceedings of an International Symposium on Environmental Impact of Radiation Releases.* Vienna, 1995.

National Research Council, Committee on the Biological Effects of Ionizing Radiations, Board on Radiation Effects Research, Commission on Life Sciences. *Health Effects of Exposure to Low Levels of Ionizing Radiation: BEIR V.* Washington, D.C.: National Academies Press, 1990.

National Research Council of the National Academies, Committee to Assess Health Risks from Exposures to Low Levels of Ionizing Radiation, Board on Radiation Effects Research Division on Earth and Life Studies. *Health Risks from Exposure to Low Levels of Ionizing Radiation: BEIR VII,* phase 2. Washington, D.C.: National Academies Press, 2006.

Petryna, Adriana. *Life Exposed: Biological Citizens After Chernobyl.* Princeton, N.J.: Princeton University Press, 2002.

Ramberg, Bennett. *Nuclear Power Plants Weapons of the Enemy,* Berkeley: University of California Press, 1984.

Scerri, Eric. *The Periodic Table, Its Story and Its Significance.* Oxford: Oxford University Press, 2006.

Schottenfeld, David, and Joseph F. Fraumeni Jr., eds. *Cancer*

Epidemiology and Prevention, 2nd ed. New York: Oxford University Press, 1996.

Solomon, Fredric, and Robert Q. Marston, eds. *The Medical Implications of Nuclear War.* Washington, D.C.: National Academy Press, 1986.

Ulsh, B. A., T. G. Hinton, J. D. Congdon, L. C. Dugan, F. W. Whicker, and J. S. Bedford. "Environmental Biodosimetry: A Biologically Relevant Tool for Ecological Risk Assessment and Biomonitoring." *Journal of Environmental Radioactivity* 66 (2003): 121–29.

Ulsh, B. A., M. C. Mühlmann, F. W. Whicker, T. G. Hinton, J. D. Congdon, and J. S. Bedford. "Chromosome Translocations in Turtles: A Biomarker in a Sentinel Animal for Ecological Dosimetry." *Radiation Research* 153 (2000): 752–59.

U.S. Environmental Protection Agency, National Service Center for Environmental Publications. "Fact Sheet on Ocean Dumping of Radioactive Waste Materials," November 20, 1980.

U.S. Nuclear Regulatory Commission. "Backgrounder on Radioactive Waste," February 24, 2011, http://www.nrc.gov/reading-rm/doc-collections/fact-sheets/radwaste.html.

Whicker, F. W. "Impacts on Plant and Animal Populations." In *Health Impacts of Large Releases of radionuclides. Ciba Foundation Symposium* 203, 74–88. Chichester: John Wiley & Sons, 1997.

Whicker, F. W. "Protection of the Environment from Ionizing Radiation: An International Perspective." In *Second International Symposium on Ionizing Radiation: Environmental Protection Approaches for Nuclear Facilities,* 136–42. Ottawa: Canadian Nuclear Safety Commission, 1999.

Whicker, F. W. "Radioecology: Relevance to Problems of the New Millennium." *Journal of Environmental Radioactivity* 50 (2000): 173–78.

Whicker, F. W., and L. Fraley Jr. "Effects of Ionizing Radiation of Terrestrial Plant Communities." *Advances in Radiation Biology* 4 (1974): 317–66.

Whicker, F. W., T. G. Hinton, et al. "Avoiding Destructive Remediation at DOE Sites." *Science* 303 (March 12, 2004): 1615–16.

INDEX

Page numbers in *italics* refer to illustrations and tables.

ILLUSTRATION CREDITS

IN-TEXT ILLUSTRATIONS

14 Penetrating Powers of Alpha Particles, Electrons, Gamma Rays, X-Rays, and Neutrons (Robert Gale)

18 Radon: Deaths Per Year (Environmental Protection Agency)

50–51 Letter from Albert Einstein (Franklin D. Roosevelt Library and Museum, Hyde Park, New York)

152 Radiation Doses from Medical Procedures (*Harvard Health Letter,* Harvard Health Publications, Harvard Medical School)

200 The Nuclear Fuel Chain (Courtesy of Ilmari Karonen and Nicolas Lardot)

COLOR IMAGES

Dispersal of Cesium-137 from Fukushima and Chernobyl (Courtesy of Marc Bocquet, CEREA)

U.S. Terrestrial Radiation (United States Geological Survey)

Lifetime Cancer Risk #1 and #2 (Courtesy of F. Owen Hoffman)

A NOTE ABOUT THE AUTHORS

Robert Peter Gale, a scientist and physician, is presently Visiting Professor of Haematology at Imperial College London. His career has focused on the biology and therapy of bone marrow and blood cancers, especially leukemias. He is the author of twenty-two medical books, and his articles have appeared in *The New York Times*, the *Los Angeles Times*, *The Washington Post*, *USA Today*, and *The Wall Street Journal*. For the last thirty years, he has led or been involved in the global medical response to nuclear and radiation accidents, including those in Fukushima and Chernobyl. He lives in Los Angeles.

Eric Lax's books include *Life and Death on Ten West*, an account of the UCLA bone marrow transplantation unit, and *Woody Allen: A Biography*, each a *New York Times* Notable Book of the Year. *The Mold in Dr. Florey's Coat*, about the development of penicillin, was a *Los Angeles Times* Best Book of the Year. He lives in Los Angeles.

A NOTE ON THE TYPE

The text of this book was set in Times New Roman, designed by the British typographer, font designer, and printing historian Stanley Morison (1889–1967) for *The Times* (London) and first introduced by that newspaper in 1932.